AF882226

Educating Leaders in Sustainability using History

Educating Leaders in Sustainability using History

By

Elizabeth Summerfield

Cambridge
Scholars
Publishing

Educating Leaders in Sustainability using History

By Elizabeth Summerfield

This book first published 2023

Cambridge Scholars Publishing

Lady Stephenson Library, Newcastle upon Tyne, NE6 2PA, UK

British Library Cataloguing in Publication Data
A catalogue record for this book is available from the British Library

Copyright © 2023 by Elizabeth Summerfield

All rights for this book reserved. No part of this book may be reproduced, stored in a retrieval system, or transmitted, in any form or by any means, electronic, mechanical, photocopying, recording or otherwise, without the prior permission of the copyright owner.

ISBN (10): 1-5275-0224-4
ISBN (13): 978-1-5275-0224-6

For my children and grandchild.

Defy the disciplines. In spite of what you majored in, or what the textbooks say, or what you think you're an expert at, follow a system wherever it leads. It will be sure to lead across traditional disciplinary lines. To understand that system, you will have to be able to learn from–while not being limited by–economists and chemists and psychologists and theologians. You will have to penetrate their jargons, integrate what they tell you, recognize what they can honestly see through their particular lenses, and discard the distortions that come from the narrowness and incompleteness of their lenses. They won't make it easy for you.

—Donella Meadows

TABLE OF CONTENTS

CHAPTER 1

ENVIRONMENTAL "WICKED" PROBLEMS

In the last two decades in Australia, there has been growing concern amongst public policy makers and leaders about the major social and environmental challenges faced by the nation.[1] These challenges contain a complexity beyond simple cause and effect relationships and require messy and incomplete solutions that involve multiple actors. These are often referred to as "wicked" problems in the management literature.[2] They are declared to be unprecedented in kind and often global in reach, requiring both national and transnational responses.[3] As a consequence of their novel and international nature, reform agendas stress the demand for innovative twenty-first century solutions.[4] The preference for management research as an evidence base for leading change in organisations and individuals can be seen in the reference lists of these policy publications.[5]

The focus on sustainability for the future can lock thinking in the conceptual realm of theory rather than grounded empiricism and, consciously or unconsciously, ignore the past as irrelevant to the solutions demanded to address current problems. Dictionary definitions of "innovation" invariably contain the word "new", the antithesis of the past or "old". This book begins from the premise that the history of environmental management has a powerful contribution to make to innovative thought and action leadership in the present for the future. It argues that the traditional focus on the new in the theory development in the field of management and organisation studies has often ignored past knowledge. Paradoxically, investigating the past has the potential to unlock new ideas that may be useful to address current and future environmental management problems.

---

[1] Australian Public Service Commission, *Tackling Wicked Problems: A Public Policy Perspective* (Canberra: Australian Public Service Commission, 2007).
[2] ibid.
[3] ibid.
[4] ibid.
[5] ibid.

The examples of recent policy papers that follow testify to the ubiquity of management literature as an evidence base, and the general absence of historical thinking. But amongst them is one that contains tentative signs of the contribution to management knowledge and education history may offer. In 2007, the Federal Government produced a discussion paper titled *Tackling Wicked Problems: A Public Policy Perspective.*[6] The Public Service Commissioner declared that adequately addressing such problems was an "evolving art" in which it was vital to recognise that no "quick fixes" or "simple solutions" were possible.[7] An important first step was "grasping the big picture, including the interrelationships among the full range of causal factors".[8] This capacity for holistic vision was seen as unfamiliar to established modes of educating leaders and citizens alike. Essential to defining new approaches to managing, these problems would be a "reassessment of some of the traditional ways of working and solving problems" that would be inclusive of all "stakeholders and citizens".[9] Critical environmental problems were named as two of four specific examples. Climate change and land degradation contained both multiple causes and effects and involved a broad spectrum of the community.[10] In relation to climate change, the Commissioner declared it to be:

> a pressing and highly complex policy issue involving multiple causal factors and high levels of disagreement about the nature of the problem and the best way to tackle it. The motivation and behaviour of individuals is a key part of the solution as is the involvement of all levels of government and a wide range of non-government organisations (NGOs).[11]

Explicit to a sustainable future was the core requirement of changing the thinking and action of individual leaders, citizens and relevant public and private sector organisations. How this was to be approached was articulated in a companion publication *Changing Behaviour: A Public Policy Perspective.*[12] Explicit too in the reference lists of both documents is the evidence base used to frame the publications, which drew almost exclusively on management and organization literature.

---

[6] ibid.
[7] ibid., iii.
[8] ibid., iii.
[9] ibid.
[10] ibid, 1.
[11] ibid, 1.
[12] ibid.

History for tackling wicked problems

In 2009 the abstractions of this policy language were shockingly grounded in the realities of a sequence of natural disasters which began on 7 February 2009 in Victoria. On this day, Australia experienced its worst bushfire in recorded history. One hundred and seventy-three people lost their lives in a firestorm that swept in one direction then another across forests under some of the most severe fire weather conditions in the world. A Royal Commission was established immediately. It served both as a symbol of the impact of the extraordinary catastrophe and a mechanism for the formal investigation of its causes and remediation. By 2010 sixty-seven recommendations had been produced with a sweep as comprehensive as the tragedy to which it responded. These included recommendations designed to effect long-term, national changes in specialist and citizen behaviour alike. For example, in Recommendation Six the Commissioners intended to leverage the innovation of the first national education curriculum to ensure subsequent generations of Australians acquired a level of environmental literacy and self-management that would keep them safe in future firestorms. The recommendation was that

> Victoria lead an initiative of the Ministerial Council for Education, Early Childhood Development and Youth Affairs to ensure that the national curriculum incorporates the history of bushfire in Australia and that existing curriculum areas such as geography, science and environmental studies include elements of bushfire education.[13]

Similarly, under the heading "Research and Evaluation" the Commissioners sought to implement a research agenda that was explicitly interdisciplinary in its reach, incorporating the sciences and social sciences, the forests, and the people who live in them:

> The Commonwealth establish a national centre for bushfire research in collaboration with other Australian jurisdictions to support pure, applied and long-term research in the physical, biological and social sciences relevant to bushfires and to promote continuing research and scholarship in related disciplines.

These two recommendations, focussed on general education and research, have since been expanded in policy responses to a series of more recent

[13] Victorian Bushfires Royal Commission, 'Final Report Recommendations', Recommendation 6. http://www.royalcommission.vic.gov.au/Assets/VBRC-Final-Report-Recommendations.pdf, 1.

environmental disasters.[14] They reflect a pervasive theme in recent environmental management literature – that of the foundational importance of "co-management" and "social learning" by all stakeholders.[15] Such research powerfully reinforces the necessity of developing flexible and adaptive thought for action amongst leaders, managers and the general public alike.

The Royal Commissioners' Recommendation Six stands out in naming history as a key discipline for affecting this change in behaviour. Management research and public policy papers that enlist their findings as an evidence base are more dependent upon those disciplines with positivist, rather than humanist, methods. Perhaps it was in subsequent deference to this methodological preference that the Commissioners' emphasis on the value of history was significantly diluted in implementation.

Aims and objectives

This book aims to explore the potential contribution of history to changing behaviour in response to environmental "wicked" problems. It argues that the accessibility of historical narratives to both lay and specialist audiences can expand the evidence base and influence theory in ways that complement the positivism of the social and environmental sciences. It does so through the case study of environmental management history, framed by management theory, in response to the call from environmental historians for a more pragmatic environmental history that actively engages with other knowledge systems.[16]

The rationale for the selection of the particular management theory and historical environmental problem for this case study follow.

---

[14] See for example, N. Dufty, "Engagement or Education?" Australian Journal of Emergency Management. 26,3 (2011): 35-39.

[15] See for example, P. Olsson, C. Falke and F. Berkes, "Adaptive Co-management for Building Resilience in Social–Ecological Systems," Environmental Management 34(1) (2004): 75-90; D. Armitage, M. Marschke and R. Plummer, "Adaptive co-management and the paradox of learning," Global Environmental Change 18(1) (2004): 86-98.

[16] S. Dovers, "Sustainability and 'Pragmatic' Environmental History", Environmental History Review, 18(3) (1994): 21-36; S. Dovers, 2008, "Can Environmental History Engage with Policy", History Australia, 5(1): 3.2-3.6.

Environmental history framed by management theory

While environmental history seeks to produce research across the disciplines, particularly of science and history, it retains a strong humanities footing.[17] Leading environmental researcher and practitioner of environmental history Stephen Dovers has argued that the latter can lead to the production of knowledge that, though of inherent interest, is not matched by the pragmatism needed if the field is to have broad impact in achieving an environmentally sustainable future.[18] In order for this to occur, he argues, environmental historians must more actively engage with other methodologies that have an accepted capacity in the current management culture of decision-making to influence change, such as management and organisation scholarship.

In 1994, Dovers' call for more "pragmatic" studies in environmental history that would make a direct contribution to sustainability research, in addition to being of inherent interest to academic historians.[19] By 2008, Dovers still believed in the capacity of environmental history to positively influence change for sustainable living.[20] Responding to the question "can environmental history help save the world?", he answered with a qualified yes. But it still needed, he said, to "better engage with modern policy debates and negotiate its interactions with other knowledge systems".[21] Dovers' argument for a more applied environmental history included his and colleague Eric Pawson's claim that history can act as an "interdiscipline".[22] Such a claim resonates strongly with the scholarship of some applied historians and historiographers.[23] It also begins to shed some

---

[17] T. Griffiths, "The Humanities and an Environmentally Sustainable Australia", Australian Humanities Review 43 (2007). http://www.australianhumanitiesreview.org/archive/Issue-December-2007/EcoHumanities/EcoGriffiths.html.

[18] Dovers, "Sustainability".

[19] ibid.

[20] ibid.

[21] ibid, 3.2.

[22] E. Pawson and S. Dovers, " Environmental History and the Challenges of Interdisciplinarity: An Antipodean Perspective", Environment and History 9, 1 (2003): 53-75.

[23] See for example, J. Appleby, L. Hunt and M. Jacob, *Telling the Truth about History* (New York: W.W. Norton & Company, 1994); H. Stretton, *The Political Sciences* (London: Routledge and K. Paul, 1969); J. Tosh, *The Pursuit of History* (London: Routledge, 2010); G. Davison, "Paradigms of public history", Australian Historical Studies 24(97): 4-15.

light on the potential value of the discipline alluded to by the Bushfire Royal Commissioners.

Similarly to Dovers' interest in applied environmental history, management scholars have debated and promoted the role of the historical method in broadening the scope and impact of management research.[24] This has been part of a larger call for greater relevance of academic research to management practice, and the need for fundamental changes to theorising.[25] In a comprehensive summary of the epistemological and ontological debates so far, Matthias Kipping and Behlul Usdiken, identify three categories for the use of history in management scholarship.[26] These are "history *to* theory", "history *in* theory", and "historical cognizance".[27] Their stated purpose is to increase the visibility and influence of history in management and organisation theory for the expanded evidence base the discipline can offer, as well as the enhanced relevance this can bring to the application of theory. In terms of their categories, this book employs that of history *to* theory. Kipping and Usdiken define this category as history serving as evidence to "develop, modify or test theories". [28]

The book will make a conceptual and empirical argument that brings together the established preference of management disciplines for theory, and the more dominant empiricism of the historical method. In doing so, it proposes one way of achieving a pragmatic environmental history that could contribute to changing the behaviour of leaders and citizens alike, and help to validate or modify management theory. Its principle purpose is to make the contribution of an exploratory case study to Dovers' plea for more pragmatic environmental history.

---

[24] See for example, G. Jones, M. van Leeuwen and S. Broadberry, "The Future of Economic, Business, and Social History", Scandinavian Economic History Review 60(3) (2012): 225-253; G. Jones and D. Wadhwani, "Schumpeter's Plea: Rediscovering History and Relevance in the Study of Entrepreneurship," Harvard Business School, Working Paper 29, (2007).
http://www.hbs.edu/faculty/Publication%20Files/06-036.pdf.

[25] R. Suddaby, C. Hardy and H. Quy, "Where are the New Theories of Organization?" Academy of Management Review 36(2) (2011): 236-248.

[26] M.Kipping and B.Usdiken, "History in Organization and Management Theory: More the Meets the Eye," The Academy of Management Annals 8(1) (2014): 535-588.

[27] ibid, 535.

[28] ibid, 536.

Method

The primary audience envisaged for the book is environmental history scholars. But, as a piece of pragmatic historical research, it also envisages an audience of environmental management scholars and practitioners. The historical narrative that forms the centrepiece of the book also aims to speak to lay audiences. This range of target audiences was developed in response to calls in the academic literature and public policy documents for including stakeholder education to achieve the "co-management" required to address complex environmental problems. Given the breadth of audience, this book offers a case study which combines analysis of relevant theory with historical narrative. The narrative analyses the leadership of the innovation of forestry in Australia (primarily) and the United States (comparatively). The comparison aims to draw generalisations about the development of successful leaders.

The management theory used to explore the contribution of history is drawn from the literature of environmental wicked problems. Systems thinking, a sub discipline of management scholarship is seen as key to the definition and management of "wicked" problems generally and environmental wicked problems specifically.[29] Within this branch of management research, leading systems thinker Peter Senge, along with colleagues Otto Scharmer, Joseph Jaworski and Betty Sue Flowers posited the Theory of the U in 2005 in *Presence: Exploring Profound Change in People, Organizations and Society*.[30] The theory proposes that it is not enough to examine the "what" and "how" of leadership, the most commonly studied areas of behaviour. Critical to understanding the quality of leadership is the much less examined who of the leader, the foundational elements of the person leading, which in turn provide a much clearer insight into the rationale for the what and how of their leadership. The theory has since been further refined by Otto Scharmer, with a particular emphasis on "leading from the emerging future", reflecting the usual temporal preference of management scholarship for the present and future

---

[29] R. Horn and R. Weber, "New Tools for Resolving Wicked Problems", Strategy Kinetics LLC (2007).
http://www.strategykinetics.com/New_Tools_For_Resolving_Wicked_Problems.pdf.

[30] P. Senge, J. Jaworski, O. Scharmer and B. Flowers, *Presence: Exploring Profound Change in People, Organizations and Society* (London: Nicholas Brealey, 2005).

rather than the past.[31] Drawing on Kipping and Usdiken's work, the research questions for the book that relate to the theory are: does an historical case study of successful management of a complex problem support the validation of the theory? Moreover, does the case study suggest modifications to the theory for further research?

The Theory of the U acts as a relevant case study of management theory for the book and is used to select and frame the questions raised by the historical evidence. The history of a complex environmental problem – that is, the innovation of the first formal Australian forest authority – is used to produce new environmental historical knowledge, and to test, and possibly modify, the Theory of the U.

Management theory and systems thinking

Since their definition in the seminal article by Horst and Rittel in 1973, the complexity of wicked problems has been recognised as containing "systems of systems".[32] Systems thinking is a sub discipline of management which is concerned to redress the traditional fragmented approach to problem solving through a holistic apprehension of problems and an iterative effort towards resolution based on continuous feedback provided by the system itself. Problems of environmental sustainability are acknowledged to be complex in nature and unresponsive to linear cause and effect treatment.[33] Such problems involve human as well as natural ecosystems, including multiple stakeholders such as agencies, managers, policy-makers, scientists, politicians and the general public. The Theory of the U was conceived with both individual and collective application in mind.

A key element of the theory is what its authors call the "blindspot" of leadership, which may be found in individuals, organisations or society as a whole.[34] Analysis of leadership, they say, has typically focussed on the "what" or the "how" of response to problems. This leaves unexplored the "who" of leadership, "the inner place or source from which we operate, both individually and collectively".[35] The ability to acquire

---

[31] O. Scharmer, "Leading from the Emerging Future", Paper presented to Federal Ministry for Economic Cooperation and Development, November 13, 2011, Berlin.

[32] Horn and Weber, "Tools for".

[33] P. Balint, R. Stewart, A. Desai, and L. Walters, *Wicked Environmental Problems: Managing Uncertainty and Conflict* (Washington: Island Press, 2011).

[34] O. Scharmer, "Uncovering the Blind Spot of Leadership", Leader to Leader, 47 (2008):52-59

[35] Senge et al., *Presence*, 5.

and access the self-knowledge as a basis for leaders to determine the "what" and the "how" of leadership, argue Senge et al., has largely been absent from research. The authors point to the existence of this wisdom in the past – more specifically, the remote ancient philosophies of Greece and China.[36] The more recent history of the Industrial Revolution, Senge argues elsewhere, has produced instead machine-like thinking and action and caused the current crisis of leadership for sustainable societies.[37] Case studies for exploring the nature of the "who" of leadership are generally limited to the recent past.

It is argued that a more rigorous approach to history linked with systems thinking, particularly the Theory of the U, can contribute to the evidence base of the theory. Furthermore, it can offer constructive models from the Industrial era to support the tackling of contemporary environmental wicked problems. The historical case studies of complex problems – that is, the introduction of scientific forestry in Australia and the United States – are drawn from the industrialising nineteenth century. In these cases, the development of the "who" of leadership is examined through their formative learning environments.

Historical case studies: leadership for introducing scientific forestry

The first historical case study begins by reconstructing a complex environmental problem of nineteenth century Australia: the diminishing forests in the colony of South Australia. The narrative of the thought leadership of parliamentary debate about forestry in the first half of the 1870s is recounted in detail in order to argue that "wicked" problems have precedents, not in their specific historical content, but in their relative complexity. Revealing that complexity not only offers evidence of precedence, but also provides an empirical case study from which constructive lessons can be learnt about how to, or how not to, think and act when confronting contemporary wicked problems.

In order to examine the "blindspot" of leadership, three key figures in Australia's first formal forest authority, the South Australian Forest Board (and the first in an independent British colony of the Commonwealth), are investigated in more detail: George Goyder, Surveyor General and Chair of the Forest Board; John Ednie Brown, first

[36] ibid, 179.

[37] P. Senge, 2012, "Creating Schools for the Future, not the Past, for All Students", Leader to Leader, 65 (2012): 44-49.

Conservator of Forests; and Boyle Travers Finniss, ex-Premier and member of the Board. The Board, the first such body in Australia, was established in 1875 and was the outcome of five years of parliamentary debate. The "who" of each man's adult leadership style is examined through the reconstruction of the ways in which they learned to think, see and act in their environments. Such consideration of the individual themselves as a complex system in a continual process of evolution is not new. Senge's research has also expanded to include the quality of education, recognising not only its formative effect on the adult but also the value of applying systems thinking principles to schools as organisations and children as prospective citizens and leaders. In doing so, Senge has been mindful of the statement by prominent twentieth-century leader of organisational quality improvement Edward Deming, that

> the prevailing system of management has destroyed our people. The destruction starts with toddlers. . . .
> The fundamental task of leadership is transformation of this system . . . [which is] the same system in education and business.[38]

Senge quotes Deming as a prelude to arguing that the industrial age produced a school system that reflected the highly mechanised and specialised forms of work demanded by societies undergoing industrialisation. The transition of societies in the twenty-first century out of the industrial age, he argues, demands new modes of learning to produce a more humane, holistic and less machine-like approach to educating future citizens and leaders.[39]

The educational histories in the book provide a comparative insight into successful and unsuccessful leadership and between successful leaders to suggest successful principles for such education. In Senge's research these have been overlooked in the large generalisations about historical epochs such as the industrial age. The value of such examples to current management research and practice is in the distillation of essential educational principles that may be applicable across the continuum from childhood to adult learning for contemporary leadership development.

In order to test the hypothesis that there was a significant connection between formative learning and successful adult leadership, the comparative case study of Gifford Pinchot's leadership is reconstructed from the historical evidence. Pinchot was instrumental in establishing the first United States Forest Service in 1905 under the presidency of

---

[38] ibid, 44.
[39] ibid.

Theodore Roosevelt, and is considered to be the "father" of US forestry. The analysis of his educational history aims to counter the methodological problem in leadership biography summarised by Geoffrey Jones and Daniel Wadhwani that "the primary drawback with such studies arises from deriving meaningful generalizations about entrepreneurship from individual cases". [40] By examining the educational influences of Gifford Pinchot and comparing these with those of George Goyder, a sounder argument can be made for the essential learning principles that shape the judgement of leaders capable of addressing wicked problems. The inclusion of Gifford Pinchot also seeks to test the hypothesis that in the industrial age, as now, the problems of forestry were transnational. While the specific historical context and kind of problem were clearly vastly different from present times, the complexity of problem formulation and resolution can be seen to have strong parallels.

The following are the research questions asked of the historical record:

What were the difficulties of introducing scientific forestry in the 1800s in Australia, and how do they resemble those of a contemporary "wicked" problem?

What did leadership of the problem look like at different stages of its evolution?

How had leaders learned early to see their environment, and how did this impact their adult leadership?

What are the similarities and differences between the education of the successful Australian leader of forestry and his USA counterpart?

What are the lessons to be learned for the leadership of present and future complex environmental problems from these case studies?

Contribution to theory and practice

This book aims to make a contribution to knowledge in two ways: by developing an exploratory case study of environmental history framed by

[40] Jones and Wadhwani, "Schumpeter's Plea".

management theory as one method for contributing to "pragmatic" environmental history; and by forming a new piece of environmental management history.

The narrative of environmental management history consciously engages historical study with other "knowledge systems" to enhance its potential application. It is not surprising that much of the literature that examines the role which history could play in management and organisation research is written by scholars trained within the field. This book, instead, approaches the combination from a predominantly historical perspective, but shapes the historical research using theories from management and organisation studies.

Specifically, the application of the Theory of the U to the historical material aims to contribute new knowledge for application across specialist environmental managers, educators, policy-makers and lay community members interested in the first principles of the what, how and who of thinking and action for sustainable living in a complex social and natural environment.

The book begins with a review of the conceptual case for history's contribution to the understanding and resolution of complex, or "wicked", environmental problems. This is followed by the construction of the historical case study. It outlines the chronology and detail of the introduction of public forestry in the South Australian parliament in 1870. In order to demonstrate the complex nature of the environmental problem the chapter does not condense the parliamentary discussions but seeks to reveal the messy protractions and convolutions of debate. The discrete stages of progression from idea to legislation to implementation of forestry are described in detail to demonstrate their messiness and protraction. The comparative educational biographies of three of the leaders of formal Australian forestry follows in order to argue the connection between the early education of each man, their consequent ability to think across disciplines and other boundaries, and their success as leaders of environmental innovation. The conclusions drawn from the Australian case study of environmental education for successful leadership are then tested through a comparison to Gifford Pinchot, the founder of forestry in the USA. The comparison acknowledges the global nature of major environmental problems, historically and currently. The comparison also demonstrates that Australian environmental history has a contribution to make to the transnational discussions of leadership development for successful management of complex environmental problems. A concluding chapter draws together the historical evidence to make the case for the contribution of history to resolving complex environmental problems. It concludes by applying the lessons learned about successful leadership of such problems to the present example of the selected recommendations of the Victorian Bushfire Royal Commission on education and research.

CHAPTER 2

ENVIRONMENTAL HISTORY AND LEADERSHIP THEORY

The previous chapter proposed that pragmatic environmental history has a contribution to make to a better understanding and resolution of today's complex environmental problems, or "wicked" problems as they have been called in the management literature. The book offers one model for the increased engagement of history with other knowledge systems, which leading Australian environmental scholar Stephen Dovers argues is essential to demonstrating the applied value of history.

The opportunity was outlined for environmental history's engagement with recent management and organisation research interested in expanding the epistemological foundation of their field by including the historical method. It outlined the call from some environmental historians for a more pragmatic history, and the view of some management and organisation scholars that history has a greater contribution to make to the inherently practical purposes of their research. It also indicated that the management sub-discipline of systems thinking, which is key to addressing complex environmental problems, has not exploited rigorous historical research for its purposes.

This chapter expands on the outline of the previous chapter to examine more closely the literature of applied history and environmental history to see what historians have said about the potential application of their discipline to the consideration of current policy and practice leadership. It does so to explore the question: is there an inherent public value already evident in the discipline? And, if so, is there a need for a more pragmatic environmental history as Dovers argues? A review of the discussions about the value of history to management and organisations research follows. This looks at the practical value of history from the perspective of a field that occupies a powerful place in providing evidence-based research for public and private leadership, including environmental leadership. It does so to explore Dovers' proposition that history engage outside its own discipline to render its findings more directly pragmatic. It asks the question: can environmental historians render their primarily humanist, empirical research more pragmatic by inhabiting the different positivist, theoretical tradition of management and organisation research?

Chapter 2 provides an expanded rationale for the selection of the systems thinking Theory of the U, and the historical evidence for reconstructing the narrative of an environmental wicked problem of the past: the leadership of the innovation of forestry. The hypothesis to be tested through this combination is that, because of the temporal focus of much systems thinking on the present for the future, the past has been largely ignored as a potential source of innovative ideas for environmental management, and that framing the research questions of the historical record by this systems thinking theory can provide new historical knowledge to more directly inform present environmental leadership, while also testing the theory.

The primary audience envisaged for this book is environmental historians of management. Chapter 2, therefore, begins by reviewing the arguments from inside the discipline of history generally, and then environmental history specifically. As a secondary audience, the book aims to speak to management and organisation scholars, including systems thinkers, sympathetic to the application of history to their research. It aims to do so principally from the perspective of the historian – that is, by demonstrating the practical insights historical data can offer systems thinking theory. The Theory of the U serves as the case study.

History as a pragmatic discipline

Some historians might take issue with the view, represented by Dovers, that history lacks pragmatism. They value history because, they argue, it offers a complementary kind of knowledge to science – one which has become devalued in relation to the positivist methods of the sciences and many of the social sciences. Historians Joyce Appleby, Lynn Hunt, and Margaret Jacob offer an historical explanation for why society has come to esteem disciplines that practise the scientific method above other disciplines, especially in searching for explanations or solutions to present problems.[1] They describe the consolidation of the Scientific Revolution through the Industrial Revolution in the late eighteenth and nineteenth centuries, and note that the reinforcement of the fragmentation of knowledge begun in the Enlightenment. Religious, affective, spiritual and

[1] Appleby et al., *Telling the Truth.*

ethical - or subjective - knowledge was segregated from, and diminished in relation to, apparently value-free, rational, material and mechanistic - or objective - knowledge.[2] Academic disciplines and organisations, which were responsible for the production of knowledge, absorbed the cultural value of esteeming the scientific perspective over other ways of thinking. Even disciplines that studied human behaviour, such as history, began to use scientific methods to try to impose order, predictability and credibility on the human subjects of their study by becoming methodologically similar to their scientific colleagues. The social sciences, they argue, increasingly derived their esteem from this epistemological association with science. But, note the authors, contemporary historians have largely eschewed such mechanistic approaches to the examination of human behaviour, seeking instead to interrogate its full range of complexity and specific contexts.

The Australian historian and public intellectual Hugh Stretton has argued that disciplines which try to emulate science for its replicable certainties must search for regularities regardless of whether or not they produce useful knowledge. Research which truly wants to understand why humans make the choices they do, he said, will look at behaviour that is not quite like anything a physical scientist studies, because it is "neither perfectly predictable nor perfectly inscrutable".[3] The researcher whose definition of rigour requires them to be "as abstract, general, objective, quantitative and unfeeling as they can" will pursue methods that "more often hinder than help the discovery and understanding of social facts".[4] Stretton asserts that the purposes and methods of the scholar of human endeavour who creates useful knowledge will be quite unlike those of the physical scientist. Their purposes are likely to include the civic, social, political and instrumental value of what they learn, and their methods will not be confined to those of the quantifier. Stretton also challenges the common view of the value-freedom of mechanistic and statistical data that aims to measure human activity. Such a view contained, he argued, a false promise of objectivity and a false credibility that derived its status from falsifying comparability to the data of inanimate objects gathered by physicists.

Appleby et al. share Stretton's view that history offers a different kind of objectivity in knowledge creation. In doing so, they note with disappointment the "enduring dichotomy" of positivism which has set up a false contest between "absolute objectivity and totally arbitrary interpretations

---

[2] ibid, 17.

[3] Stretton, *Political Sciences*, 164.

[4] ibid, 159.

of the world of objects".[5] Rigour, in this context, is always measured in terms of scientific method and outcome. They instead suggest the adoption of a "practical realism" when interpreting human behaviour. Such realism prizes rigour while acknowledging the inescapable interplay of subjectivity and objectivity in the construction of genuine knowledge about human beings, who are clearly of more complex construction than the physical objects able to be studied by the scientific method.[6] In the interplay between man and nature, which is the basis of environmental history, the arguments of these historians offer a valuable warning to remember the distinction between the two subjects of research. They also argue that the pragmatism of the discipline of history derives, from its exploration and elucidation of the complexity, the contingency and the contextual detail of phenomena. If pragmatism is to be assessed by the generation of replicable laws of theoretical constructs that seek to define and predict human behaviour on a large scale, as is the case with economics or the physical sciences, then history will not qualify – nor will decision-makers or leaders have access to knowledge that could make a practical difference to problem identification and solution.

In addition to those historians who defend the discipline's intrinsic, even practical, value to making meaning of events, there are applied historians who, as their name suggests, seek to make more direct connections between the past and the present for the future. Applied historian John Tosh takes Appleby's concept of "practical realism" a step further in arguing the value of history when applied to contemporary policy matters. He stresses the inherently democratic nature of historical knowledge, its accessibility to a range of audiences, and its potential to engage citizens and professionals alike to influence change. Tosh regrets that, "time and again, complex policy issues are placed before the public [or public managers] without adequate explanation of how they have come to assume their present shape, and without any hint of the possibilities which are disclosed by the record of the past".[7] A deeper understanding of history, argues Tosh, has an inherent contribution to make to rich and meaningful knowledge construction for thought leadership and action.

More specifically, Tosh argues, the contribution of "practical realism" through the historical method offers an "inventory of alternatives" where we can find "patterns of thinking or behaviour that are immediately accessible to us" and from which contemporary generalisations may be posited. The preferred technical approaches to social and political

---

[5] Appleby et al., *Telling the Truth*, 247-251.

[6] ibid.

[7] Tosh, *The Pursuit*, 10.

problems, he says, tend to compartmentalise human experience "into boxes marked 'economics', or 'social policy'…each with its own technical lore". What is needed instead for more purposeful leadership is "openness to the way in which human experience constantly breaks out of these categories"[8]. The lateral links to be found between the compartments are much easier to spot with the benefit of hindsight rather than the immersion of the researcher in what some management scholars call the current "messes" they seek to interpret.[9] Tosh notes that

> more and more historians are now investigating themes of topical relevance. They do so not as a propaganda exercise, but in the conviction that there are valuable insights to be learnt from the findings of historical scholarship…. If society looks to historians for "answers" in the sense of firm predictions and unequivocal generalizations, it will be disappointed. What will emerge from the pursuit of "relevance" is something less tangible but in the long run more valuable – a surer sense of the possibilities latent in our present.[10]

Foregoing universally applicable laws of prediction, therefore, does not mean an acceptance that no human event is open to understanding, generalisation and leadership.

Such apology for the contribution history can make, often expressed by comparison with the superior rigour typically attached the scientific method, itself has a history of over a half century. In 1961 Edward Carr agreed, in a much-read book which is still in print, that scholars of society in the first half of the twentieth century had "consciously or unconsciously desiring to assert the scientific status of their studies, adopted the same language and believed themselves to be following the same procedure".[11] They had subsequently realised that "the human being is on any view the most complex, natural entity known to us, and the study of his behaviour may well involve difficulties different in kind from those confronting the physical scientist".[12] In 1981, reviewing a pioneering piece of Australian environmental history, Hugh Stretton again summarised the key features of that difference. Historical insight offers the capacity to see wholes as well as parts and to be explicitly critical of the

---

[8] ibid, 12.
[9] R. Ackoff, "The Art and Science of Mess Management", Interfaces, 11:1, (1981): 20.
[10] Tosh, *The Pursuit*, 334.
[11] E.H. Carr, *What is History?* (Camberwell: Penguin Group, 1961): 58.
[12] Carr, *What is history?*, 69.

values that influence behaviour in particular contexts, including those of the historians themselves.[13]

The views of historians on the philosophy and practical value of history to the public good undoubtedly resonate strongly within the discipline in which shared training makes them appear self-evident. However, are they shared by practitioners of the newer sub-discipline of environmental history?

Ideas of environmental history's pragmatism

Environmental historians hold a range of views on the appropriate audiences for, and practical application of, their sub-discipline. As we have seen Dovers, an environmental scholar, has both used and valued historical methods, but has also been frustrated by the field's failure to seek to have a more direct impact on the pressing environmental sustainability problems of contemporary society. More pragmatic environmental history, he argues, would support the necessary dramatic changes needed in public and private environmental leadership. The "notion of pragmatic environmental history", he says, is that which "as well as being of inherent interest and furthering the discipline", serves a social as well as academic purpose. Such history also aims to make a "positive and practical contribution to environmental management and the quest for ecological sustainability".[14] Dovers wants to see a more direct outreach of historians to practitioners of other "knowledge systems" in order for the discipline's potential contribution to be understood and heard beyond those trained to see and appreciate its relevance.

Other environmental historians see the practical value of the field in its public intellectualism. In the example of the 2009 Black Saturday fires cited earlier, Australia's and the United States' environmental historians made a public contribution to the early thought leadership aimed to influence future change.[15] Tom Griffiths noted the paradox of the wealth of practical scientific and emergency management knowledge that had accumulated in Australia since the devastating fires of 1939. However, he also noted the quarantining of that knowledge inside the specialist silos that had generated it and the lack of a holistic management capacity to aggregate those silos for the understanding and solving of the large and

[13] Davison, "Paradigms", 15.

[14] S. Dovers, "Sustainability and 'Pragmatic' Environmental History: A Note from Australia," *Environmental History Review*, 18(3) (1994): 21-36.

[15] T. Griffiths, "We have still not lived long enough," Inside Story, February 14 2009, http://inside.org.au/we-have-still-not-lived-long-enough/.

complex physical and social problems the fires made evident. Griffiths then pointed to the extraordinary amount of expert knowledge Australia had acquired since 1939 and our apparently equally mounting failure to act on it.[16] We had seen nothing less than a "revolution in scientific research and environmental understanding", he said. We knew the fire ecology of the forests with unprecedented scientific insight. In fact, we "knew that this terrible day would come". Why, then, had we been powerless to prevent it? Some of the answer, he argued, lay in the segregation of that specialist knowledge from other complementary sources of knowledge, and from public and management practice. Useful knowledge had become isolated inside parts of research or government institutions, lost to its other parts and to a wider public audience that badly needed it rendered in accessible terms for nothing less than their own survival. Part of the correction of the contradiction, Griffiths asserted, lay in all Australians being brought to a greater historical knowledge of their forests and climate, and its impact on their thinking and action in the natural environment. Griffiths argued that the empowerment of the public through access to specialist knowledge, made intelligible by the universality of historical narrative, was key to preventing the recurrence of disaster on this scale. He saw the same potential for improved holistic, or systemic, understanding amongst those specialists who were responsible for creating the knowledge of forests and climate needed to underpin public education, policy and management practice. Historical narrative could deliver these stories to encourage a change in behaviour, too.

American fire historian Stephen Pyne added his voice to the local chorus of bewilderment about Australians' failure to learn from the past.[17] The fires were "a horror", he agreed, "even by Australian standards, which is saying much". But there should have been no mystery about their cause: "Australia is a fire continent: it is built to burn", he declared simply.[18] His bewilderment came from incomprehension about why the Australian public, policy-makers and environmental managers had forgotten this, and thus failed to plan and act accordingly. His hope was that there could be the re-development of much greater awareness of forests and how they function in the Australian climate, and a deep reconsideration of how, as Australians, we live with and act in them.

While Griffiths and Pyne share with Dovers a view of the contribution environmental history can make to changing the thinking and behaviour of specialist leaders and the general public, Griffiths and Pyne

---

[16] ibid.
[17] S. Pyne, "Black Saturday: the Sequel," Peeling Back the Bark, 10 February 2009.
[18] ibid.

place greater value on the persuasive power of historical narrative. These scholars trust the universal accessibility of narrative to carry that change. In a compelling argument for the power of narrative, Griffiths argues that "narrative is not just a means, it is a rigorous and demanding one". While the traditional scientific method separates causes from one another, he says, testing them individually in turn, "narrative … carries multiple causes along together". In doing so, he continues, "it enacts connectivity".[19] Griffiths is keen to promote the "rapprochement" of science and history because "we need both methods" and perhaps because the combination of both will have a more powerful and balanced impact on public leadership. But prominent American environmental historian William Cronon doubts the value of this pursuit.

Cronon has posed the direct question: what are the uses of environmental history?[20] He points to the insights that a reconstruction of the environmental past offers the present, especially for his own most prized audience, the general public. Echoing Griffiths, he believes that historical narrative contains the democratic potential to reach and inform a citizenry who ultimately decide as voters the environmental policies and practices they want pursued by governments, their policy makers and managers. The principal obstacle he sees in attempting to address ahistorical professional audiences is the superior place the positivist methods have assumed in their professional training. They would first need, he said, to be persuaded of the rigour and value that can be found in historical narrative as a means of influencing their practical thinking.[21] He prefers not to spend his time as an apologist for history at the expense of writing for a general public for whom such a defence is unnecessary. In Appleby's terms, he believes they already appreciate as inherently interesting and instructive the "practical realism" historical narrative contains.

But the case Cronon makes for the combined rigour and accessibility of historical narrative for his main audience contains the same potential to influence the other audiences he names – scientists, policy makers, politicians, social scientists and environmental managers – as long as they are open to innovative ways of making meaning of their environments. Story, argues Griffiths, "is actually a piece of disciplined magic, of highly refined science. It is the most powerful educational tool we possess . . . allowing for multiplicity and complexity at the same time

---

[19] T. Griffiths, "The Humanities".

[20] W. Cronon, "The Uses of Environmental History," Environmental History Review, Fall (1993): 1-22.

[21] ibid, 5.

as guaranteeing memorability".[22] In the universality of its mode, narrative has the ability to transcend the specialisms of training and be instructive across disciplinary borders. However, Cronon is more sceptical. Perhaps, like Dovers, he regards the argument of the universal accessibility of narrative in a management knowledge culture where esteem attaches to the scientific method, as an unduly optimistic assumption of history's potential for influence.

As a practitioner of various disciplines, Dovers is at once hopeful of and sceptical about the prospects for history's utility in debates about environmental change leadership. He has successfully employed environmental history as one of several disciplines in his integrated investigations of environmental policy and management. For him and many of his colleagues, the primary audiences are the specialist decision-makers Cronon does not seek to engage, and Griffiths seeks to engage through an interdisciplinary mix of science and history. Dovers locates the power for significant change in politicians, policy-makers, environmental managers and researchers, as well as the general public. These audiences are less likely to have historical training, and to appreciate the self-evident connections of past to present named by historians. Dovers is driven by the urgent need for thought and action leadership in environmental change. In an article for *The Ecological Society of America* Fischer, Fazey, Gross, Dovers and Ehrlich argue that the priority areas for ensuring environmental sustainability require nothing less than "reforming formal institutions, strengthening the institutions of civil society, improving citizen engagement, curbing consumption and population growth, addressing social justice issues, and reflecting on value and belief systems"[23]. The ambition of this transnational, cross-sectoral and inter- and intra-organisational brief, argue the authors, demands an approach that is of necessity interdisciplinary with direct application to problem solving. Elsewhere, Dovers makes the case with Eric Pawson for history as a pivotal discipline in this interdisciplinary mix. They argue that,

> of all substantive foci, past uses of environments and their future sustainability have generated greater quantity and diversity of interdisciplinary ventures than any other, and so offer a source of much needed project narratives, intersections and analyses of interdisciplinary engagement. With more elaborate engagement, environmental history,

---

[22] Griffiths, "The Humanities".

[23] J. Fischer, R. Dyball, I. Fazey, C. Gross, S. Dovers, P. Ehrlich, R. Brulee, C. Christensen, R. Borden et al., "Human Behavior and Sustainability", Frontiers in Ecology and the Environment, 10 (2012): 153.

> arguably the environmental "interdiscipline" that attracts the greatest disciplinary variety, may not only improve its own explanations, but become the leading laboratory in the interdisciplinary experiment.[24]

Dovers and Pawson also see the broad accessibility of narratives of environmental history as having the potential to influence not only the general public, but also specialist audiences. They see history as a disciplinary key for translating the less readily accessible knowledge and language of environmental specialists into the accessible language of narrative. They see the potential of historical narrative as a basis for creating a shared understanding of the complexity of problems and as a vehicle for advancing disciplinary collaboration for environmental problem solving, but they seek a more deliberate and immediate engagement of history with other disciplines for this purpose than the other historians cited here.

In broad terms, historians, including applied and environmental historians, share a view about the capacity of their discipline to have a practical impact on social outcomes. History's ability to reconstruct wholes rather than parts, to develop evidence-based narratives that reveal and cohere complexity, that bring together the lived reality of subjectivity and objectivity, of valuing, and context in the rigorous construction of knowledge are common themes of the literature. But is this enough to persuade those outside the discipline and particularly those in the management disciplines who, as Davison and Tosh, argue have the preferred ear of public leadership and do not necessarily see rigour in humanist disciplines? Dovers, as critical friend of environmental history, argues the view that more explicit outreach to other "knowledge systems", which have a more established audience of public leadership, is required. For him, the "practical realism" needed is one that seeks the compromise of integrating historical methods with these methods, rather than arguing for the stand-alone complementary value of history. Given the ubiquity of management research and practice in the public leadership of environmental problems, this field is an obvious one with which to attempt the integration Dovers argues for.

The following section looks more closely at the views of some management and organisation scholars who also believe it is time for their field to be paying more attention to history. This discussion provides the rationale for the selection of a management theory for leadership and the framing of a case study that aims to demonstrate one form of the explicit engagement for which Dovers argues.

---

[24] Pawson and Dovers, "Environmental History," 69.

Management and organisation research: one vehicle for more pragmatic environmental history

A recurring complaint of some management and organisation scholars is that current research has too readily assumed the superiority of scientific method in the generation of useful knowledge.[25] Dan Remenyi, for example, thinks that "numerical based research, sometimes referred to as the positivistic approach, plays a very important role, if not a dominant role, in business and management studies research". Echoing historians Appleby, Hunt and Jacob, he traces this dominance to the influence of the Scientific and Industrial Revolutions.[26] In response to these epistemological concerns, Michael Rowlinson and Charles Booth note a growing call from management and organization theorists for an "historic turn" as one way of restoring balance to the way knowledge is constructed.[27] Echoing the historians cited above, they, too, point to the fallacy of trying to understand the complexities of human activity in leadership or institutions by looking for laws of behaviour that replicate those of machines. Similarly, Mayer Zald sees historical thinking as an effective challenge to the typical model of research adopted from the physical sciences, which he argues is too "universalist and presentist" in "the search for general and abstracted laws" of human enterprise.[28]

The work of Joseph Schumpeter from over a half-century ago has inspired other scholars to ask what an historical approach can contribute to the study of entrepreneurship in business and management studies.[29] In a review of research since Schumpeter's call for "creative destruction" to drive entrepreneurship in the 1940s, Geoffrey Jones and Daniel Wadhwani trace the demise of historical scholarship in favour of behavioural and cognitive approaches. Studies using such approaches, they believe, lack the historical and social context of earlier studies, dismissed in an effort to seek universally applicable laws of human and organisational behaviour. They identify two fallacies in the context-free generalisations typical of

---

[25] See for example, M. Zald, 2002, "Spinning Disciplines: Critical Management Studies in the Context of the Transformation of Management Education", Organization, 9: 365–85; D. Remenyi, "Tell Me a Story – a Way to Knowledge", The Electronic Journal of Business Research Methodology 3 (2005): 133-140.
[26] Remenyi, "Tell Me", 133.
[27] C. Booth and M. Rowlinson, "Management and Organizational History: Prospects", Management & Organizational History 1(2006): 5-30.
[28] Zald, "Spinning Disciplines", 381.
[29] J. Schumpeter, "The Creative Response in Economic History", Journal of Economic History 7 (1947): 149-157.

the scientific approach: they assume they are applicable to all times and places; and they assume entrepreneurial behaviour is “fundamentally new or *sui generis*”.[30] The dominance of the scientific method in management and organization studies, they regret, has left a legacy in the culture of scholarship that continues to limit innovation. In a different twist on the notion of “practical realism”, the authors note the “veritable impossibility” of publication in leading management journals if they do not use what they call “standardized social science methodology”. Pragmatic historical research and writing may not be very practical for advancing an academic career outside of history. The obsession with research drawing on “large statistical databases”, they say, and the exclusion of “‘non-rigorous’ studies using small samples or qualitative data” has been “hugely distorting” of entrepreneurship research, but is self-perpetuating in this publishing culture.[31]

On a more positive note, Wadhwani and Marcelo Bucheli, argue in *Organizations in Time: History, Theory and Methods* that it is timely to capture the current trend in management scholarship towards history by taking steps to make “historical analysis theoretically relevant for the questions asked by management and organisation scholars”.[32] Wadhwani and Jones demonstrate the enrichment of current management theories of entrepreneurship using an historical approach.[33] Roy Suddaby argues that an historical approach takes into account “the complexity of internal and external factors” which offers a “less clean”, if not “messier”, picture, but a “more accurate and realistic” one that considers individuals and an ever-changing context.[34] These scholars share an interest in bringing together management research’s embedded culture of theory with the complexities and contingencies of empirically based historical narrative.

In a meta-analysis of recent literature, Matthias Kipping and Behlul Usdiken identify three categories in which management scholars currently employ history. They name these “history *to* theory”, “history *in* theory” and “historical cognizance”.[35] Common to each of the categories, and crucial to Kipping and Usdiken’s argument, is how the use of history relates to theory. Like Wadhwani and Bucheli, they name the central importance of theory to the “scientization” of management culture since

---

[30] Jones and Wadhwani, “Schumpeter’s Plea.”
[31] ibid., 28.
[32] M.Bucheli and D. Wadhwani, “A Return to the Past: History and Organization Studies”, Aegis 10(1) (2014): 5-13.
[33] ibid, 10.
[34] ibid, 9.
[35] Kipping and Usdiken, “History in Organization”, 535.

the mid-twentieth century, and the necessity of invoking it if history is to have a more credible voice within knowledge systems committed to theorising.[36] In this, they share Jones' "practical realism" for marrying history's humanist narrative method with the positivist preference for theory in order to generate a cross disciplinary dialogue for the expansion of knowledge in both disciplines. They define "history to theory" as the use of historical data to "develop new or modify extant theories".[37] The use of the past as an integral part of the construction of the theoretical model itself they call "history in theory". In each of these uses of history, the resulting theory was seen to be universally applicable and timeless. But, where authors were seen to be taking history most seriously, that is they were "aware of and explicitly considered the limits to generalizability resulting from the use of history", they were seen as exercising "historical cognizance". They were, in other words, acknowledging the contingency and complexity of data that qualified their theorising and not necessarily seeing the theory as universally applicable.

The recent interest in management and organisation theorists in expanding their field through engagement with history offers an opportunity to test the development of a more pragmatic environmental history through the use of one or more of the categories identified in the Usdiken and Kipping's comprehensive summary. The book will employ the category "history in theory".

The following section articulates the rationale for the choice of systems thinking Theory of the U.

Choosing a relevant management theory

Complex problems of environmental sustainability are often classified as "wicked" in the current policy and management literature. Conversely, the term "tame" is applied to problems that are simpler, or those that are linear in cause and effect with more immediately accessible formulations and solutions. Wicked problems elude both simple definition and resolution. Horst Rittel and Melvin Webber are attributed with having first coined the term "wicked" problem in 1973.[38] Since then, the management scholarship has used descriptors such as "complicated and complex", "social messes", and "tightly interconnected, economically, socially, politically,

[36] ibid.

[37] ibid, 536.

[38] H. Rittel and M. Webber, "Dilemmas in a General Theory of Planning", Policy Sciences, 4 (1973): 155-169.

technologically", to describe their multi-faceted nature.[39] So prominent has the scholarship of wicked problems become that the Australian Public Service Commission published a report in 2007, with the title "Tackling Wicked Problems", and indicated that many of government's major policy problems needed to be first conceived of and then approached framed by this field of management scholarship.[40] A companion piece called "Changing Behaviour", published in the same year, elaborates on the thought leadership and action demanded of public leaders and the general public alike.[41] The literature confirms that fundamental to addressing such problems is enlisting the scholarship and practice of systems thinking, a sub-discipline of management and organisation studies.[42]

Russell Ackoff, prominent management scholar and systems thinker, outlines the transformation required by public leaders in conceiving of and addressing complex problems using systems thinking.[43] He begins with a definition of the term. "Systemic thinking", he ventured, is "holistic" rather than "reductionistic". It demands a "synthetic" rather than "analytic" approach to problem identification and solution. Reductionistic and analytic thinking were features of scientific or positivist disciplines that seek to determine the properties of wholes from the properties of their parts. Conversely, he explained, holistic and synthetic thinking "derive properties of parts from properties of the whole that contains them". In general, he argued, public (and private) sector leaders had been trained to think about problems in ways that meant they did not understand that improvement in the "performance of parts of a system taken separately may not, and usually does not, improve performance of the system as a whole". The definitions of other leading systems thinkers share many of Ackoff's descriptors. Paradoxically, in view of the under-utilisation of history by systems thinkers, they also resonate strongly with the earlier descriptors of history.

In 2013, Ockie Bosch, Nam Nguyen and Daowei Sun reported the outcomes of three major projects in which they identified and

---

[39] R. Ackoff, "The Art and Science"; R. Horn, "Knowledge Mapping for Complex Social Messes", A presentation to the "Foundations in the Knowledge Economy" at the David and Lucile Packard Foundation, July 16, 2001; Australian Public Policy Commission*, Tackling.*

[40] ibid.

[41] Australian Public Policy Commission (APPC)*, Tackling*.

[42] Balint et al., *Environmental Wicked Problems*, 2011; J. Ogilvy, "Systems Theory, Arrogant and Humble, World Futures", The Journal of Global Education, 69, 4-6 (2013): 332-344.

[43] Ackoff, "The Art and Science", 20.

intervened to resolve complex social problems. The title of their article in the journal *Business Systems Review* declared the essential need for innovative thinking and action in their approach. They defined "new ways of thinking" as synonymous with systems thinking. This, in turn, could be described, they said, as "a holistic and integrative way of appreciating all the major dimensions of a complex problem", which then "enables the formation of effective management strategies (systemic interventions) with long lasting outcomes".[44] Similarly Kevin Ronan, proposing the desirable thought and action demanded by "wicked" problems, reiterates the need to analyse the problem from a "holistic viewpoint", coming to an understanding of such problems in "systemic, interconnected and non-linear terms".[45] It is not surprising, given the emphasis on innovative thought leadership in systems thinking, that education of the young, as well as the older, on embedding a systemic approach is an element of much of the literature.

As mentioned, the resonance of the definition of systems thinking with the literature of historians describing their discipline is strong. But, with limited exceptions, systems thinking is not a field of management scholarship which has engaged strongly with historical research as an evidence base for the validation or modification of its theory. It is, however, a field powerfully interested in issues of environmental sustainability and leadership.

Systems thinking for environmental leadership

Systems thinkers have joined the search for new ways of generating knowledge for environmental sustainability. Old ways, they argue, are limited by the assumption that knowledge systems for environmental problems are linear, predictable, and measurable, and thus subject to manipulation with the kinds of knowledge that the scientific method provides. New and useful knowledge about environmental complex systems, the chief characteristic of which is "irreducible uncertainty", cannot be reductionist. Generating this knowledge, as Donella Meadows argues, also demands the total engagement of our individual and collective human system of understanding phenomena. It involves internal as well as

[44] O. Bosch, N. Nguyen and D. Sun, "Addressing the critical need for a 'new way of thinking' in dealing with complex issues facing our societies", Business Systems Review 2(2) (2013): 48-70.

[45] K. Ronan, "Solving 'wicked problems' linked to disasters, risk and uncertainty – children are truly our future", Australian Journal of Emergency Management 29(1) (2014): 8-9.

external knowledge. Although trained as a scientist of computer systems, she had come to mistrust complete dependence of her rational mind for understanding and acting on problems of sustainability. Her matured view was that nothing less than "our full humanity – our rationality, our ability to sort out truth from falsehood, our intuition, our compassion, our vision, and our morality" was required in meeting the challenges of a sustainable world.[46] Fellow systems thinkers Senge and his colleagues also warn of the danger of an over-dependence on scientific methods, a habit he sees as a by-product of the industrial revolution of the nineteenth and twentieth centuries.[47] Echoing the views of the historians and management scholars cited earlier, they believe that future challenges require the exploration of new ways of knowing and acting in the world. They believe that

> as long as our thinking is governed by habit – notably by industrial, "machine age" concepts such as control, predictability, standardization, and "faster is better" – we will continue to re-create institutions as they have been, despite their disharmony with the larger world, and the need of all living systems to evolve.[48]

Getting to know systems, their exponents advise, requires patience and humility. Observers must watch, listen and experiment. Meadows warns against the learned assumption in the technological age of immediate response and action. Patient observation, she argued, "discourages the common and distracting tendency… to define a problem not by the system's actual behaviour, but by the lack of our favourite solution". If we watch systems over time, we will come to understand that "self-organizing, nonlinear, feedback systems are inherently unpredictable", such that unthinking intervention can "destroy the system's self-maintenance capacities". But this sort of watchfulness also creates opportunities for limited intervention, as we discover the "triggering events" of change in complex systems, and that means paying attention "to what is important, not just what is measurable". Echoing fellow systems thinkers, she reinforces the demand to observe whole systems rather than confining ourselves to discrete and apparently manageable bits of them. Leading systems thinking for problem-solving means, Meadows writes, aiming "to enhance total systems properties, such as creativity, stability,

---

[46] D. Meadows, "Dancing with Systems", Systems Thinker, 13 (2002): 2-6.
[47] Senge, *The Necessary Revolution*, 9.
[48] P. Senge, O. Scharmer, J. Jaworski and B. Flowers, "Awakening Faith in an Alternative Future", Reflections: the SoL Journal on Knowledge, Learning and Change (2004) 7: 5.

diversity, resilience, and sustainability – whether they are easily measured or not". Whole systems likely to be understood while managers or organisations are locked inside disciplinary silos. In terms which resonate with the argument of environmental historians Dovers and Pawson, she declares that problem solving demands a willingness to be led "across traditional disciplinary lines" to "admit ignorance and be willing to be taught, by each other and by the system".[49]

Greater knowledge of leaders' own human systems is a corollary of seeking greater understanding of external systems. Deeper self-knowledge is an indispensable but under-explored condition of holistic appreciation of the external environments that surround us. Senge and Scharmer argue that the "largely unexplored territory" in management and leadership research "concerns not the what and how . . . but the who: who we are and the inner place or source from which we operate, both individually and collectively". This, they argue, is the "blindspot" of management research in leading significant change. Much more represented in research has been the "what" and the "how" of such leadership. In their most significant challenge to dominant modes of thought, they look to an examination of the "who" of management thought and practice to effect "profound change in people, organizations and society".[50] Leadership, they argue, needs to be shared across the traditional boundaries of specialist managers and individual citizens for sustainability to be effective.

While not explicitly engaging with historical methods, systems thinkers do see the value to their work of the economist's notion of path dependence, in which decisions made in the past and under different circumstances continue to limit our decisions in the present.[51] This is evident in Senge's and Meadows' description of the industrial and scientific revolutions as the source of current problems of thought leadership and management practice. Senge in particular has written extensively on education.[52] He explicitly connects the system of education and its foundations in the industrial age with the perpetuation of thinking that is reductionist and management structures that are of the command and control variety. He points to the advent of mass education coinciding with that of the industrial age. In order to fit the first to the second,

---

[49] Meadows, "Dancing".
[50] Senge et al., Presence, 5.
[51] See, for example, J. Forrester, "System Dynamics, Systems Thinking, and Soft OR", Systems Dynamic Review 10 (1994): 245-56.
[52] P. Senge, N. Cambron-McCabe, T. Lucas, B. Smith, J. Dutton, and A. Kleiner, *Schools That Learn: A Fifth Discipline Fieldbook for Educators, Parents, and Everyone Who Cares About Education* (Boston: Nicholas Brealey, 2000).

children were schooled in modes and methods that moulded them for the military-like hierarchies of the factory floor and corporation. Such organisations required submission to institutionalised power structures and their siloed operations. Schools were modelled not only these power structures, but also the siloed dispensation of disciplinary knowledge.

In order to prevent the obstruction of deep understanding through this fragmentation of knowledge and begin to produce a different type of citizen and leader for the twenty-first century, Senge argues, learning needs to stem from the lived experience of children and adults. To prevent the persistence of arbitrary power, including knowledge construction for the approval of established authority, students must be encouraged to assume greater responsibility for their own learning. Such an approach to learning – that is, learning in wholes rather than parts – then becomes hard-wired. Senge invokes the pioneer of the quality improvement movement in post-World War II America, Edward Deming, to make explicit the importance of systems thinking and learning in schools for the production of future citizens and leaders.[53] He quotes Deming, who declared that "the prevailing system of management has destroyed our people" and that "the destruction starts with toddlers", so that the "fundamental task of leadership is transformation of the system" which, declares Deming, is "the same system in education and business".[54] While Senge argues the important causal connection between the educational history of the child and an adult capacity for leadership at an individual and collective level, he does not engage with what Kipping and Usdiken call "historical cognizance".

Senge and some of his colleagues employ history largely to point to the causal nature of present problems or, more distantly, to times and cultures which contain evidence of preferred modes of thought and action.[55] Scharmer is entirely dismissive of history. He argues that a focus on the past is not very helpful, that it can impede thought leadership for innovation. Neither view of history allows for identifying the complexities of human endeavour in the past and the constructive evidence for thought leadership to be found there. This book argues that systems thinkers have yet to exploit the complementary insights and opportunities of a holistic view of history and the historical method to enhance the insights systems thinking can offer complex problem definition and resolution. As Griffiths declares, "not only do we have to advance the "frontier", we also have to remember what we think we already know, and we have to give new

[53] Senge, "Creating Schools", 44.
[54] ibid.
[55] Scharmer, "Leading from".

meaning to what is forgotten".[56] The history of the complex environmental problem of the innovation of forestry in the industrial and scientific era offers an opportunity to examine a systems thinking theory, devised by Senge, Scharmer and colleagues, engaging with the possibilities of historical case study. In the language of Usdiken and Kipping, the book presents a case of "history in theory".

Systems thinking: Theory of the U

In 2005, Senge, Scharmer, Jaworski and Flowers posited the Theory of the U to chart the process that effective leaders should move through in decision-making and action to arrive at the best possible outcomes.[57] It poses the following fundamental questions for leaders: who am I? What is the relationship between this self-knowledge and the work I need to do? Beyond the ability to see and think across boundaries in the external environment-of disciplines, of organisations, of cultures, of politics-these scholars theorised the necessity of taking the complexities of the external system and considering them in ways that integrate external and internal environments. The Theory of the U describes the critical importance of leaders taking external information and processing it reflexively through their understanding of who they are in order to make genuine meaning of and act in their circumstances. As a result of the pressures of time, busyness and expectations for immediate outcomes, argue Senge et al., leaders often do not read and process information deeply but rather respond at the more superficial level of decision-making and action, resulting in less effective and thus sustainable consequences. Short-term gains may be achieved but in the longer term further problems become manifest as the result of a lack of holistic engagement with the entirety of the problem, and what the leader can bring to its resolution. Scharmer and Senge argue the integral importance to leadership of the "who" that sits at the base of the U and forms the "blindspot" of leadership research, education and practice.

Scharmer has since refined the theory, calling it Theory U, making its unmistakable core the interior process of leading.[58] It is in this later research that Scharmer declared explicitly the irrelevance of a temporal focus on the past. He concluded that, in opposition to a leader's

[56] Griffiths, "The Humanities".

[57] Senge et al., 2005.

[58] O. Scharmer, *Theory U: Leading from the Future as it Emerges* (San Francisco: Berrett-Koehler, 2009).

natural deference to past learning and practice, they should instead focus on the emerging future and lead from that basis. Scharmer described his findings from interviews with over 150 "world's thought leaders and innovators" to Germany's Federal Ministry for Economic Cooperation and Development. He reported two distinct types of learning by leaders: those who learned from the past; and those who learned "from the future as it emerges".[59] He noted that while it was natural to do the first, "sometimes the past is not very helpful", that the "experiences of the past are themselves obstacles to coming up with new ideas".[60] In addition to this future focus, he found that the success of an innovative leader depended on the "inner place" from which they operated. Scharmer's first conclusion does not strictly discount the investigation of innovative leaders of the past. But its stress on the present for the future tends to reinforce the "presentism" which those proposing the use of history in management and organisation scholarship suggest is one of its limitations. This book proposes that the past contains evidence that can expand upon the theory by offering historical case studies of innovative leadership from "an emerging future" and of the "inner place" from which that occurred.

The book does not suggest that historical case studies offer a superior alternative to present-day case studies. Rather, that they can enrich these studies by opening up hitherto unexplored areas of evidence which can be repositories of innovative thinking for the still largely ahistorical field of management research. Collecting qualitative data in the present does not have the vantage point of hindsight, which can be particularly valuable when issues of sustainable innovation are being considered. Historical narrative has a beginning, a middle and an end. It offers a whole story rather than the necessarily less complete stories of the present.

Conclusion

Systems thinking occupies a position of vital importance in the interrogation and resolution of environmentally wicked problems. An expansion of its evidence base for thought leadership, while testing the more positive role history could play in its theorising aims, could make an important contribution to knowledge for the leadership of complex environmental problems. Sitting alongside this is the hypothesis that combining management and organisation theory and historical narrative

[59] Scharmer,"Leading from".
[60] ibid.

can provide a substantive contribution to a more pragmatic environmental history, while also constructing the "practical realism" of a narrative of leadership learning for the general public. The first element of this hypothesis is presented in the next chapter.

CHAPTER 3

INNOVATING AUSTRALIAN FORESTRY: A 19TH-CENTURY WICKED PROBLEM

This chapter constructs the narrative of a complex environmental problem of the past: the innovation of Australian forestry. The problem has been chosen because it conforms to the description of a complex, or "wicked", environmental problem, as described by the management literature. How was the colony to supply a growing demand for timber as a primary resource for sustainable settlement in a landscape subject to the driest climate on the continent? How was it to do so while also ensuring that the developing agricultural economy and ability to feed a growing population were coincidentally fostered? How was the land to be settled in ways that provided the infrastructure necessary to realising the ideals of an envisaged civic society? And how was it to do both when the knowledge of the country and the climate was itself very new? Such questions surrounded the innovation of forestry, rendering it a "wicked" nineteenth-century problem. Its complexity resonates strongly with the current descriptors of such problems, including having "many interdependencies"; the possibility of "unforeseen consequences" that could arise in addressing it; having "no clear solution" because of the multiplicity of social, environmental, economic and political factors involved; being "socially complex" because of the range of those potentially affected; crossing organisational responsibilities of central and local governments; and involving changing the thought and action of politicians, policy makers and general public alike.

The primary conceptual purpose of this chapter is to demonstrate that the complexity of environmental problems is demonstrable in the past, and that historical case studies can therefore add empirical evidence that can support a clearer understanding of what such problems look like and how their complexity might be addressed.[1] The primary empirical purpose of the chapter is to reconstruct from the archival material the thought and

[1] APPC, *Tackling*.

action leadership exercised by parliament as a whole, and by Goyder and other key figures of forestry's innovation in particular.

The historical evidence consists principally of the Hansard records of South Australian Parliamentary Proceedings and Debates for the period 1870 to 1875. Newspaper articles for the same period provide supplementary information of the reactions of journalists and the general public to the parliamentary process and to the questions surrounding the introduction of forestry. However, the primary focus of this chapter is to describe the relationship between the champions of forestry and the political leadership of the day as it played out within the multi-faceted nature of debates. It is acknowledged that to reveal the full complexity of the forestry's legislative innovation other primary sources, such as records of government correspondence or further newspaper articles, could also be interrogated. But the purpose of this chapter is to establish the existence and nature of a complex problem of environmental innovation in the past and the beginnings of its thought leadership. This provides the context in which the problem and leadership then played out. The Hansard record provides a comprehensive and highly reliable source of evidence for this purpose.

The narrative that follows begins with an overview of the South Australian social and planning context in which George Goyder introduced the concept of forestry as part of a larger philosophy of planned settlement of the colony. This philosophy was his own, but sat on a continuum with the social reform ideals of the colony's founders. The legislative progress of forestry follows in three sections that correspond to significant milestones in a protracted process. The deliberately detailed description aims to reveal the often excruciating course of an environmental wicked problem. It does so in order to add real human detail to the often disembodied abstractions of current policy descriptors which assume a lack of precedent and therefore a lack of empirical data. The description is structured to demonstrate, in recreated real time, the messy chronology of parliamentary debate, action and inaction. It aims to show the slow movement of progress toward decision-making typically rendered in the writing of research findings in a much reduced, rational form. However, to do this here would be to disguise rather than elaborate the protracted, often emotive, sometimes trivial nature of the thought leadership exercised by parliamentarians in the gradual momentum towards the realisation of formal forestry in the colony. The writing is framed by the same principle used by systems thinkers Senge et al. as they conceived the writing of *Presence: Exploring Profound Change in People,*

Organizations and Society.[2] As the clarity of their theoretical thinking began to take shape, they acknowledged the usual tendency of writers about leadership, organisations, and social change to "spare readers the messiness and uncertainty of their journey".[3] In doing so, not only is the rational mess disguised, they argued, but also the emotional and experiential. If this pretence of logical, linear sequence is maintained in the formulation of ideas, then it contributes to the notion of a simple, rather than complex, problem. This complexity of the environmental innovation of forestry, the messiness that characterised its circular, protracted progress through parliament from 1870 to 1875, is impossible to portray authentically without the deliberate articulation of much of its chronology.

The first section of the narrative from 1870 to '71 covers the introduction to parliament of the ideas of and need for innovation of public forestry in South Australia. The second section describes the frustrations of progress and the eventual willingness of the primary parliamentary advocate, Friedrich Krichauff, to settle for a more modest legislative outcome, an Act to encourage the planting of forest trees, agreed in 1873. This section also indicates how the aims of the original bill to initiate large scale public forestry on public lands surveyed for the purpose were continued quietly in the background of a sympathetic legislative process. A final section covers the introduction of a bill, which led to the passing of the Forest Board Act in 1875, and the creation of the first formal forest service in Australia, and in an independent colony of the British Commonwealth.

The South Australian Planning Context

South Australia was founded on philosophical principles. The colony was itself an innovation which provided fertile ground for subsequent innovative thought and endeavours. Before outlining the trajectory of the forestry debates, some background in the social and philosophical foundation of South Australia provides the cultural context in which Goyder and Krichauff introduced the issue.

South Australia was founded as a British colony in 1836. It was distinctive amongst its fellow Australian colonies as a convict-free

---

[2] Senge et al., *Presence*.
[3] ibid, 17.

settlement, planned against philosophical principles.[4] While other Australian colonies, such as neighbouring Victoria, also demonstrated progressive policies in their settlement, South Australia had the opportunity to conceive from its inception the ideals of founding a new civic settlement. One tenet of this foundation, a persuasive element for Goyder, was likely to have been the distinctive absence of a state religion.[5]

Amongst the ideas of the planned settlement, notes geographer Michael Williams, was the importance of nature to British colonial settlements. Urban parklands were a feature of Australian and New Zealand colonial planning.[6] These colonies were founded in the midst of rapid European industrialisation in the late eighteenth and early nineteenth centuries, with associated concerns about the potential damage to the physical, social, moral and aesthetic environment considered. Public parklands in South Australia's planned system of settlement were a uniquely defining component of the capital city Adelaide's identity, forming a belt of vegetation enclosing its commercial centre.[7] The parklands were one reflection of an applied theory of systematic colonisation articulated by Edward Gibbon Wakefield and realised in the early survey of the colony by Colonel William Light. Careful planning of the physical environment reflected the founders' aspirations for the balanced social, economic and environmental settlement of those emigrating from an overcrowded Britain, in which the deleterious effects of slums were becoming increasingly evident. That is not to say that South Australia was entirely exceptional in their planned management of the landscape; Joseph Powell points also to the role of key public servants in Victoria, Thomas Mitchell and Baron Von Mueller, in exercising a similar "landscape authorship".[8] But the colony was unique in having a planning philosophy that framed its settlement under the *South Australian Colonisation Act 1834.*

Williams suggests the connections between early nineteenth-century British social reformers and the National Colonisation Society that

---

[4] M. Williams, "George Woodroffe Goyder: a Practical Geographer", Proceedings of the Royal Geographical Society of Australasia (S.A. Branch) Inc. (1978): 79.
[5] D. Pike, *Paradise of Dissent: South Australia 1829-1857* (Melbourne: Melbourne University Press,1957).
[6] M. Williams. "The Parkland Towns of Australia and New Zealand", Geographical Review 56(1) (1966): 67-89.
[7] J. Sheldrick, *Nature's Line: George Goyder, Surveyor, Environmentalist, Visionary* (Adelaide: Wakefield Press, 2013), 41.
[8] J. Powell, *An Historical Geography of Modern Australia* (Cambridge: Cambridge University Press, 1988) xv.

devised South Australia's settlement program.[9] Robert Owen's concept of a model factory and community settlement, established in New Lanark, Scotland, was the likely prototype for Adelaide, he posits, and for the two hundred or so South Australian townships (that Goyder later ensured mimicked its design). Owen's concept of a self-supporting, cooperative community was to be laid out as a quadrangle containing housing and community facilities such as parks, playgrounds, schools, meeting halls, and infirmaries, and separated from factories and workshops by a substantial belt of parklands[10]. In his 1966 research of the conceptual origins of South Australia's design, Williams acknowledged the impossibility of identifying direct connections between such reformers as Owen and the planners of South Australia. (A direct link between Owen and Goyder is established later). But he did establish clear links between the reformist movement and the philosophical principles of the colony's foundation. The colony, he said, was founded as part of the search for an ideal form of settlement. This included the utilitarian philosophy of "promoting the health and happiness of the greatest number" and occupied many of the "educational, political, and economic thinkers of the time".[11] Systematic colonization, and consequently the settlement of South Australia, was born of this movement. Williams noted that "there was hardly a man among the colony's founders who was not in the forefront of some reformist movement in Britain in the early nineteenth century". He points to the links between these men. Jeremy Bentham was a utilitarian philosopher, a business partner of Owen's (1813-1829) and a prominent member of the National Colonisation Society for South Australia. Colonel Robert Torrens was an economist, and a member of the Committee of Inquiry that looked into Owen's "Plan for Villages of Co-operation" in 1819. He became chairman of the National Colonisation Society in 1835. Owen was a friend of Robert Gouger, another prominent member of the Society.[12] The belt of parklands surrounding the township of Adelaide was to be repeated in over two hundred smaller townships in South Australia alone, largely during the planning oversight of Goyder.[13] Williams demonstrated the influence of the "practical geographer" George Goyder in the longevity of this design feature, and suggested the Surveyor General's early education may have been Influential his determination to

[9] Williams, "Parkland Towns", 2.
[10] ibid., 3.
[11] ibid.
[12] ibid, 2.
[13] ibid, 1.

sustain this planning feature.[14] The book will provide evidence in a later chapter that confirms Williams' speculation. The early history of the planned foundation of the colony, which included survey of land before distribution, encouraged forward thinking, including considerations about the timber needs of a growing population. Such consideration was also encouraged by the physical landscape of the colony. It was, and is, the driest state in the driest inhabited country in the world.

Certainly necessity, to some extent, bred the invention of forestry legislation and forest management in the colony. Afforestation and conservation were essential to a sustainable settlement. Historians and geographers have argued that environmental necessity alone explains the innovation of forestry management in South Australia.[15] Also of significance were the social ideals of a civil society that had guided the colony's foundation. The sections that follow describe the diversity of utilitarian and philosophical motivations that framed the first five years of parliamentary debates about forestry. These were not simply debates about subsistence but contained issues that indicated scientific, social, economic and political concerns about a sustainable settlement.[16] Goyder's practical foresight, as well as his philosophical sympathies with the colony's foundations, were vital factors in the innovation of Australian forestry in the colony.

Goyder in the South Australian Context

The philosophical context of South Australian settlement offers an initial insight into the connections between Goyder's background, his attraction to the colony and his powerful views about civic settlement. Goyder arrived in Australia in 1848 at the age of 22. He lived at first with his sister and brother-in-law, a Swedenborgian minister, in Sydney. But by 1851 he was resident in Adelaide, joined the Department of Lands in 1853 and rose

---

[14] Williams, "Practical Geographer".

[15] I. Tyrrell, *True Gardens of the Gods: Californian-Australian Environmental Reform, 1860-1930* (Berkeley: University of California Press, 1999); J. Husmann, "Transplantations: a Comparative History of Afforestation in Nebraska and South Australia 1870s- 1940s. PhD diss. (2004). The University of Nebraska, Lincoln.

[16] G. Allen and E. Gould, "Complexity, wickedness and public forests", Journal of Forestry 84(4) (1986): 20-24; M. Carroll, K. Blatner, P. Cohn and T. Morgan,"Managing Fire Danger in the Forests of the US Inland Northwest: A Classic 'Wicked Problem' in Public Land Policy", Journal of Forestry 105(5) (2007): 239-244.

in quick succession to become the colony's Surveyor-General by 1861.[17] Goyder was probably further encouraged by the colony's commitment to a non-state religion, to its embrace of dissenting denominations. Colonists from dissenting churches were actively encouraged to immigrate and continue in the practice of their faiths without impediment. Other Australian colonies had, by contrast, inherited the Church of England as a state religion. As we will see later, it was not incidental to Goyder's approach to the settlement of the land that he was a practising member of the Swedenborgian, or New Church, and that he had been raised by parents who actively lived the spiritual principles of that faith.[18] It is no surprise that the colony attracted social reformers from unorthodox faiths, such as George Goyder. In fact, the first congregation of the New Church in Australia was established in 1844 in Adelaide.[19] Goyder's background and its importance to his leadership of forestry will be expanded later.

But Goyder's commitment to the founding principles of a civic settlement framed his view of forestry and its role in sustainable settlement. The history of South Australian forestry usually begins with the speeches introducing the topic in the houses of parliament by Krichauff and John Hogkiss.[20] But defining events occurred before forestry was formally introduced to parliament in September 1870. In April of that year Goyder reported to parliament on an exploratory visit he had made to the neighbouring colony of Victoria to investigate its land acts of the 1860s.[21] Victoria enjoyed a larger population than South Australia. While this could be partly explained by the discovery of extensive goldfields in the early 1850s, Goyder sought information about the allocation of its land to determine whether there were any lessons applicable to South Australia, which could enhance its appeal to prospective settlers.[22] The environmental understanding he had developed over more than a decade's observation of his own colony, enabled him to make insightful observations about Victoria's physical environment. He reported that "the soil of Victoria does not strike me as being of much better quality, if so good, as a large portion" of South Australia. But he

---

[17] "Goyder, George Woodroffe (1826-1898)", Australian Dictionary of Biography, National Centre of Biography (1969).

[18] Pike, *Paradise*.

[19] D. Hilliard, "Unorthodox Christianity in South Australia", History Australia, 2(2) (2005): 38-3.

[20] N. Lewis, *A Hundred Years of State Forestry: South Australia: 1875-1975* (Adelaide: Woods and Forests Department, 1975) .

[21] *Register*, "Mr. Goyder's Report on the Victorian Land Act", 28 April 1870.

[22] ibid.

knew that soil was of secondary importance to climate and rainfall, and that it was "to this alone that the superiority of yield in Victoria is due".[23] But he spent considerably more space in the report summarising his views on the social disadvantages of the sequence of lands acts promulgated in the neighbouring colony.[24] These judgements demonstrate his integration of the physical and social in his evaluation, and the ethic of civic settlement that underpinned his environmental assessment. Apart from the potential damage to the natural environment that uncontrolled selection of the land could bring, he declared that

> [in] free selection before survey there are other evils of a social character, which it is perhaps unnecessary to do more than allude to in such a report as this, but which will press themselves upon a Government determining the mode of settlement of a country. Whatever truth there may be in the statements frequently made that in another colony isolated selection has been taken advantage of to facilitate and protect crime, there can be no doubt that the settlement of a country in this detached manner has a tendency to place the selectors beyond the better influences of society.[25]

Goyder's principles of settlement were deeply equitable.[26] In his role as Surveyor General, he modified and extended the Wakefieldian principles of systematic colonisation on which land distribution in South Australia was made with a pragmatic vision that Wakefield could not possibly have had because he had never experienced the reality of the colonisation he planned.[27] At their core, these principles required the prior survey of government land to ensure viable and equitable settlement of a population large enough to render the colony sustainable.[28] So, while he found much to commend in the evolution of the Victorian land laws over the previous decade he could not agree with their central tenets. He was particularly critical of "an Act which enables a few persons to pick up choice blocks of land before survey, to the detriment of the remainder".[29] It was for Goyder a matter of principle as well as simple common sense in colonies whose economies were agricultural to "divide the whole

[23] ibid.
[24] ibid.
[25] ibid.
[26] G. Goyder, *Report by the Surveyor-General upon the Disposal of Public Lands of South Australia* (Adelaide: Government Printer, 1890) 41.
[27] E. Wakefield, *A Letter from Sydney* (Letchworth: Temple Press, 1829).
[28] *Register*, 28 April 1870.
[29] ibid.

country", declaring those parts that were fitted for the purpose of pasturing only from the rest and to exclude any land from "any occupation save that for which is it is really adapted".[30] Beyond these pragmatic principles of apportioning land according to its suitability for activity, was the equity of land access for the whole community. "Selection before survey in an indiscriminate manner" advised Goyder "must be, to a certain extent, hurtful to the community, as privileges necessary to its welfare are apt to be overlooked or forgotten".[31] By contrast, selection of land following survey meant such omissions or oversights were minimised because "the necessity of essential and advantageous reserves is brought prominently forward before the Government by the officers effecting the respective surveys…and fully known to the public". He was critical of the lack of schedules in the new Victorian land bills that specified the "nature, extent and locality of reserves". [32] Such specification was not only necessary for the productive use of land for agriculture, but for the proper settlement of communities throughout the colony, he reiterated.[33]

Goyder's view was that "settlement of a country in this detached manner has a tendency to place the selectors beyond the better influences of society".[34] Most important, he believed, was the prospect of damage done to the individual child and the colony as a whole by the lack of access to schools for children's education. For parents busy trying to render their land productive "labour is of the first importance" which means the "distance and expense of schooling is a sufficient excuse…for education being dispensed with altogether". [35] He acknowledged the benefits of "hardy men and women" raised as practical and capable workers, but felt this result was also one producing "muscle without a proportionate expanse of mind". For a man from whom the disinterested language of public administration might have been a reasonable expectation, Goyder's forceful declaration of this as a "growing evil" to be "deplored" showed the depth of philosophical commitment to his role in developing both productive and sustainable community settlement.[36] Yet, with typical balance, he went on to say that he knew planned settlement could never be sure of producing ideal communities. However gradual sale or lease of land from town centres would mean that the establishment of

[30] ibid.
[31] ibid.
[32] ibid.
[33] ibid.
[34] ibid.
[35] ibid.
[36] ibid.

schools as a centrepiece of civilised life was more likely. With the importance of his own education and his own parents' life-work in teaching, he knew from experience the truth of asserting the "inestimable advantage which education confers" and that those without it were likely to be "less useful members of society, and enjoy a less rational life". An educated population, he said, was as important to the development of a country as the labour provided by an adequate population. Goyder's holistic philosophy of sustainability and its link to education was clear. In practical terms, it was the want of education that "conduces to bad farming and the utter exhaustion of the soil".[37] The practical and the ideal, the individual and the collective, the present and future, and the natural and the human were not dualisms for Goyder, but all elements of a single whole.

The visit to Victoria had been preliminary to Goyder's drafting of a "Memo on Amendment of Waste Land Laws" to the Commissioner of Crown Lands and Immigration in preparation for discussion in the Parliament of the fate of the "wastelands" (unoccupied lands) of South Australia.[38] The resulting bill was comprehensive in its prescriptions of what was needed for the continued settlement of South Australia as a viable economy and civil society. Part of the picture was the provision of forest reserves at points around the country where they could be expected to grow to provide for the needs of communities of prospective settlers in sustainable ways and, possibly to provide an additional industry and revenue source for the colony.[39] While Krichauff or Hodgkiss were the parliamentary spokesmen for forestry's introduction, Goyder preceded and informed the debate on forestry, placing it in the larger context of sustainable civic settlement.

Following the initiation of legislation for forestry in September 1870, Goyder produced a brief report for the parliament in October declaring the best sites for reserves and stating that they should be four square miles in size.[40] To provide for settlers and to fall within the necessary climatic conditions for growth, there would be two in the northern districts, five in the south eastern and one in each of Yorke Peninsula, Port Lincoln and in all temporary reserves dedicated to the use

---

[37] ibid.

[38] *South Australian Parliamentary Proceedings* (*SAPP*), 1870-71, no. 20, 79.

[39] ibid.

[40] *South Australian Parliamentary Debates (SAPD)*, 7 September 1870, "State Forests", 706; *SAPP,* 1870-71, "Report on Suggested Forest Reserves", no. 144, 1-2.

of travelling stock to market in Adelaide.[41] Both this report and the much more expansive one produced after his visit to Victoria testify to Goyder's capacity to read problems whole, across the boundaries of social, physical, economic and environmental categories. The reader gets the strong sense that he knew what needed to be done about the innovation of forestry from the outset; it was to him self-evident. But he also knew that integral to the success of any innovation was what would now be called in the policy literature the need "to continue to focus on effectively engaging stakeholders and citizens in understanding the relevant issues and in involving them in identifying possible solutions".[42] His 1870 report indicates that he had the answers to the problems of deforestation and afforestation in ways that would continue the Wakefieldian philosophy of settling a civil society. But that was not enough. It was to take another twelve years of effort to get back to where Goyder had begun in 1870 because of the need to persuade others to the same view.

In the story of South Australia's innovation of forestry, Goyder can be situated squarely in the tradition of the founding philosophies of the colony. He was a pragmatic idealist, a social reformer like his father and, like Owen and Bentham, had the practical and political skill to bring the first principles of that same reformist zeal to the settling of Australia. Goyder was well equipped practically and philosophically to manage the ongoing settlement of the colony, including its afforestation, because of this political, social and educational inheritance, themes that will be developed in subsequent chapters. He played a significant but necessarily less vocal role in the parliamentary debates on forestry that follow.

---

41 ibid.

42 APPC, *Tackling*, ii.

CHAPTER 4

LEADERSHIP OF FORESTRY LEGISLATION

This section begins to trace the chronology of the ideas of forestry and its management in the South Australian parliament from uncertainty toward less uncertainty. A reasonable expectation of the combination of the colony's pressing need for timber, the planning philosophy of South Australia, and the guidance of Goyder might have been that formalising the innovation of forestry would be expeditious. Indeed, there was a general view by parliamentarians that the introduction of forestry was important to the colony's sustainability. But, as the debates demonstrate, there was also a willingness by members in both houses to allow discussion about this general good to be lost sight of in the contest over the detail. This included the economics of forestry as a potential revenue-raising industry, the potential of forests to effect climate change, the relative claims of private landholders and public land use for forests, and the appropriate roles of centralised and decentralised governance of forestry. The entrenched views of parliamentarians on either side of these elements of the debates helps to explain the inertia of legislative progress, and its first success at the end of the period as a much diluted Act. In the end, it took a forceful belief in forestry shared by Goyder and two other leading figures in the colony to maintain the momentum towards a tentative first step in its innovation between 1870 and 1873.

Goyder's fellow champions of forestry were, at this time, parliamentarian and German-trained botanist Friedrich Krichauff and his compatriot Richard Schomburgk, Director of the Botanic Garden. Krichauff was to be the voice that would drive the cause of forestry inside parliament. The strident leadership of these men was essential – and part of the reason for the protraction of debates. All three men shared interests in arboriculture and horticulture, but Krichauff's and Schomburgk's Germanic backgrounds shaped their views rather differently from Goyder's Scottish-English upbringing. They had arrived together in South Australia in the late 1840s as part of a large immigration from the Germanic territory of Schleswig-Holstein. They were both graduates of the University of Kiel and had begun their new lives in South Australia as a farming partnership.[1] Unsurprisingly, their views on the importance and management of forests were strongly influenced by their German heritage and their specialist training. Germany sat alongside France as the centre of old world forestry. [2] This European background may help explain why Krichauff's and Schomburgk's advocacy of forestry depended heavily on the role forests could play in transformational climate change, and therefore the agricultural economy, by dramatically increasing the rainfall of the state. Goyder did not share these views. Nevertheless, their three-way alliance on behalf of forestry was vital to its advocacy. In 1870, their scientific differences were less important than the larger issue of social sustainability. Yet, in spite of their shared commitment, the generalised concern about deforestation, and a history of social innovation that framed the colony's settlement, the early phase of parliamentary debates highlighted the challenging nature of the problem of innovating forestry.

Debates begin: will they ever end?

On 6 September 1870, John Hodgkiss moved in the South Australian Upper House "that in the opinion of this Council it is desirable to encourage, by legislative action, the planting of forest trees generally, but particularly on lands near main lines of road".[3] The next day, Krichauff moved in the House of Assembly, or Lower House of Parliament,

> that a Select Committee be appointed to enquire into and report as to what is the best size of reserves for forest purposes, and where they are to be made, to recommend the best and most economical means of preserving the native timber thereon, and of planting or replanting the reserves as permanent State forests: and what are the most valuable indigenous or foreign timber trees, having in view as well a supply for public purposes, as also an annual revenue from the sale of surplus timber.[4]

The boldness of these initiatives may be seen in the light of Michael Williams' and Joseph Powell's declaration of how fundamental to the creation of a viable society the powerful contest for arable land in

---

[1] D. Morris, "Mueller, Sir Ferdinand Jakob Heinrich von (1825–1896)", Australian Dictionary of Biography, National Centre of Biography (1974).

[2] J. Dargavel and E. Johann, *Science and Hope: a Forest History,* (Cambridge: White Horse Press, 2013) 230.

[3] *SAPD*, "Planting of Forest Trees", 6 September 1870, 1870-71, 685.

Australia was at the time.[5] Settlers wanted land for farming, and could have been expected to object strongly to any alternative use of such land. Goyder's earlier memo on the wastelands indicated the amount of viable land available for immediate survey and settlement was 2,000,000 acres.[6] These lands were situated in the area south of what had become known as Goyder's line, a highly contentious demarcation of the area to the north which unreliable rainfall made agriculture unsustainable. The limits of agricultural settlement in the colony become obvious when these two million acres are understood to be less than one-hundredth of the total area of South Australia, 245,329,920 acres, and less than one-tenth of the arable land, which totalled 26,201,600 acres.[7] Of this, only 16,427,240 acres remained available to be considered under Goyder's proposed amendment. In this picture of relative scarcity, it begins to become clearer why the section of the memo declaring that there should be acquisition of land by government for forest reserves, could be contentious. But after noting his exclusion from the calculation of land of inferior quality, Goyder noted, as an apparently self-evident assumption, the similar exclusion of "lands requisite as forest or timber reserves" so that these "should again be systematically planted – as is done in many places – in order that the supply may be kept up".[8]

Hodgkiss' and Krichauff's speeches followed a month after Goyder had simply declared the environmental necessity of approval by parliament of 320,000 acres identified for forestry purposes. The protracted parliamentary debates that ensued suggest that Goyder's memo may have been a hopeful, tactical understatement.

Hodgkiss and Krichauff could not have orchestrated a better coverage of the global breadth of the arguments for forestry. Hodgkiss led with the functional and primarily climatic reasons for allocating sixteen per cent of the available public land to tree planting.[9] Hodgkiss claimed that professional men of science around the world endorsed the view that climatic changes were produced by forests. The next day, in the other house, Krichauff raised a dramatic combination of spiritual and economic concerns.[10] He concentrated on the benefits of preservation and afforestation to the beauty of the country and its appeal to the prospective immigrants.

---

[4] *SAPD*, "State Forests", 7 September 1870, 1870-71, 765.
[5] *SAPP,* 4 August 1870.
[6] *SAPP*, 4 August 1870.
[7] ibid.
[8] *SAPD*, 6 September 1870.
[9] *SAPD*, 7 September 1870.
[10] ibid.

Hodgkiss noted the international consensus about the general value of forests to a society, but he went on to defend his motion primarily on the grounds that the planting of trees would improve the colony's climate. It was his belief that the weight of evidence demonstrated that "considerable influence was exerted over the climate by the number of trees planted in any particular district".[11] Invoking the immediate esteem that attached to scientific opinion, he cited Professor John Hutton Balfour, Chair of Botany at the University of Edinburgh, to substantiate his belief. Balfour, explained Hodgkiss, described how "the exhalation of watery fluid from the leaves of plants" influences the climate of a country.[12] He went on to enlarge the global base of his evidence citing Greece, the Cape Verde Islands and India as places where the rapid deterioration of climate and impoverished water supplies had followed the felling of forests. He offered further scientific expertise from Dr Hugh Cleghorn of India, and Dr von Mueller, the highly regarded Director of the Victorian Botanic Gardens.[13] These men of science, he said, confirmed the "influence of forests in a physical, economical and hygienic point of view [as] deserving of a more complete investigation than it has yet received". In a remark that seemed to equate forests with crops, Hodgkiss noted that "the cultivation of grain, which has so completely transformed one part of the wilderness of Australia, has already a most beneficial influence on the increase of rain".[14] Returning to more local scientific opinion, Hodgkiss cited Schomburgk's recent paper, which not only confirmed Balfour's findings, but added the weight of longitudinal scientific experimentation in "dry country where the rain only fell for a few days in the year". Here, the results were overwhelming: "the rainfall through planting of trees had increased 200 per cent".[15]

Towards the end of his lengthy and rather confusing apology for the motion, Hodgkiss made passing reference to fellow parliamentarians likely to be more persuaded by economic arguments. Considerable funds from the sale of timber could be anticipated for the Government coffers, he said, from investing in forests planted and harvested on public lands.[16] Farmers too could enjoy higher quality crop production through the bird life that would be increased by the planting of trees. They would feed on the insect population, which, without these natural predators, were sure to

---

[11] ibid.

[12] ibid.

[13] D. Morris, "Mueller".

[14] *SAPD*, 6 September 1870.

[15] ibid.

[16] ibid.

ravage the crops. Finally, Hodgkiss argued that the benefit to travellers, currently exposed in their journeying to the extremities of heat and direct sunlight, of the provision of shade must not to go unnoticed by members in their consideration of the motion.[17]

Once the motion was introduced, it was then opened up to the House for debate. William Parkin raised the matter of sustainable settlement, declaring that this measure was owed by the present generation of decision-makers and the general public they represented to the next. He, for one, did not want future generations thinking that "those who lived in 1870 were a set of fools…and neglected their duty to posterity".[18] But, as significant a concern as a sustainable future might be, Parkin acknowledged the final impotence of the Upper House to do other than affirm the problem. They lacked any authority to allocate funds for the work and so could not reasonably formulate a plan to promote it. This remained the province and the responsibility of the Lower House - but he, at least, wanted to stress his view that the matter of forestry was an onerous one appropriate for the serious deliberation of his colleagues.[19]

Returning the debate to the issue of climate, John Barrow reminded members of the still uncertain nature of the relationship between tree planting and rainfall, and considered a decision based largely on the prospect of climate change unwarranted by the inconclusive meteorological science. Instead, he offered economic and aesthetic justifications. It was indisputable that trees increased the beauty of the scenery, which in turn increased the value of property to landowners, he said. Tree production and reproduction by private landowners also provided a local source of timber at reduced cost. The alternative was the prohibitively expensive search for timber growing "on the tops of mountains and in almost inaccessible gullies".[20] Barrow had shifted the argument not only from the climatic to the economic and aesthetic, but from the public to the private sector, and from the collective to individual responsibility. He urged consideration in the new land bill of Hodgkiss' idea of the planting of trees by government along the roadways, and possibly railway lines, as well as the treatment of tree planting by landowners as capital improvement. District councils, rather than central government, should have responsibility in local planting, and private citizens could contribute the

---

[17] ibid.
[18] ibid.
[19] ibid.
[20] ibid.

wisdom of their local experience to determine which species would grow best in their region.[21]

So far, there had been no dissenting voice, just differing justifications. But, as if to ensure genuine debate, the pastoralist John Bakernow declared his opposition to the motion.[22] Preservation, not afforestation, was the government's priority, he argued. While he would lend his support in principle to the planting of forest trees, he did not agree that they ought to be planted along the roadsides if they were going to cause all of this rain to fall. He drew members' attention to the case of English roadsides where it "was the invariable practice to destroy the trees on the sides of the road" in order to "reduce the rainfall…on the roads".[23] Apparently, Baker needed no further persuasion that rainfall followed the trees. His introduction of the preservation issue would be taken up in earnest by others later. If Baker's had been a lone voice of minimal dissent, the following day's motion in the Lower House, where budgetary responsibility lay, would challenge the previous day's apparent simple acceptance of the principle of introducing forestry.

On 7 September, Krichauff moved that a Select Committee "report as to what is the best size of reserves for forest purposes".[24] In his introductory remarks, he, like Goyder, assumed broad support for the measure. Perhaps this assumption came from the confidence engendered by his combination of scientific credentials and the application of this knowledge over twenty years of farming in the colony. But his opening remarks concentrated on the spiritual rather than functional importance of trees to a new society. In a dramatic introduction, he used the words of seventeenth-century English writer, gardener and diarist John Evelyn, to evoke the symbolism forests contain:

> in a word, and to speak a bold and noble truth, trees and woods have twice saved the whole world – first by the ark, then by the cross – making full amends for the evil of the fruit tree in Paradise by that which was borne on the tree in Golgotha.[25]

He continued in a tone of ironic understatement acknowledging that he "was somewhat acquainted with arboriculture".[26] With that scientific

---

[21] ibid.
[22] ibid.
[23] ibid.
[24] *SAPD*, 7 September 1870.
[25] ibid.
[26] ibid.

credential noted, he drew parallels between the original sin of Adam and Eve and the "great evil" of indiscriminate felling by private persons of timber on public lands, especially true considering South Australia was so little endowed with forest.[27] His appeal moved from biblical precedent to sixteenth-century English history and the instructions delivered to the commanders of the Spanish Armada by their rulers. In 1588, they were charged with the task of felling the entire Forest of Dean, "knowing what an injury it would be to England if her trees were destroyed". In the following century, when domestic destruction of trees threatened the very existence of State forests, laws had been enacted that carried extreme financial penalties. In Germany, too, there were centuries-old laws that ensured sustainable forest tree planting. Residents of Frankfurt, for example, had to plant three young trees for every one felled. Even in the "barren neighbourhood" of Berlin, there existed one hundred year-old mulberry trees and "but for these the country would not have produced the quantity of silk it had done".[28] The moving oratory of Krichauff's speech combined spiritual, aesthetic, cultural and economic levers to persuade members towards his heartfelt conviction of the value of forests in South Australia.

With the large philosophical claims made, Krichauff's speech took a stronger personal and pragmatic turn. His homeland had witnessed the long-term damage of failure to sustain their abundant forests. Centuries ago, these had provided a lavish cover of vegetation. Belated recognition of the problem by government officials at the end of the eighteenth century had restored not only an impressive seventy-one-thousand acres of forest to the region, he went on, but had yielded a net income of nineteen-thousand pounds per year. A similar tale of the profit-generating capacity of trees came from Scotland. Krichauff cited James Brown in his role as Forester at the Scottish Estate of Arniston. Brown's vast experience of managing forests in Scotland demonstrated, argued Krichauff, that after sixty years a mature forest would "pay the proprietor nearly three times the sum of money that would be received by him if the land had been let".[29]

Krichauff was fully aware of an obvious objection to his reasoning: that the circumstances of, and arguments from, old world societies were not directly transposable to a young colony in the southern hemisphere. So, he invoked evidence to argue a local case from the debit side of the ledger. His investigations had shown that during the last financial year the colony had imported timber at a total cost of 68,038

[27] ibid.
[28] ibid.
[29] ibid.

pounds. The amount for the preceding year had been 81,990 pounds. Set against this was the exportation of a mere 5 pounds' worth of timber and a somewhat healthier 11,547 pounds' worth of bark.[30] To rectify this imbalance, he argued for the immediate introduction of fast-growing coniferous trees so that the colony could follow the exporting example set by the European countries he had cited. Maintaining the momentum of lessons to be learnt by the colony from European governments, Krichauff moved swiftly to recommend the planting of no less than one-thousand-acre portions of land. This should begin in the dry country to the north to maximise the benefit to climate and soil improvement, he said, quite persuaded of the climate change argument that rain would follow the forests. Ferdinand Mueller had assured him, he said, not only of the value of plantations of pine or soft timber, but the wisdom of following their Victorian neighbours' example of realising this planting in localities under the governance of District Councils.

In conclusion, Krichauff made explicit the fundamental importance of trained professional foresters to the enterprise. Only scientific men "who understood arboriculture" with credentials similar to his own, von Mueller's or James Brown's, could hope to realise the success of the initiative.[31] General support for Krichauff's motion followed. The surveyor and farmer Arthur Lindsay lent his support for quite straightforward reasons. He simply did not want the colony to suffer the fate of "some of those old countries that were formerly well timbered, but which were now barren deserts".[32] With no doubt in his mind of the self-evident value of the legislation, he then moved immediately to the practical suggestion that "jarrah or some other kind of gum-tree should be planted along their railway lines, as was done in Belgium" so that there would be "no difficulty in supplying them with sleepers when they were required".[33] Others followed in simple agreement that the European examples, particularly of the Germans, were worth emulating.

Politicians seek expert advice

The only real challenge seemed to be of a financial nature. Krichauff had moved for a Select Committee, presumably to ensure that a small number of men of arboricultural expertise were charged with informing the

[30] ibid.
[31] ibid.
[32] ibid.
[33] ibid.

legislation by bringing back to the house precise, scientific recommendations. In spite of the strong economic arguments advanced by Krichauff, John Hart, the Treasurer and Premier, counselled against an expensive committee.[34] His advice was to charge Schomburgk with the task, and to offer some increase to his allowance as compensation. Krichauff diplomatically objected that the task was too burdensome for a single man to do well and with the necessary speed. It seems likely that he also wanted to secure Goyder's voice. In the end, reports were prepared by both men, Goyder having already documented much of the evidence necessary for the bill in his earlier report. The repetition or reinforcement of issues and options necessary to secure environmental innovation was becoming an evident part of the public process.

Meanwhile, the upper house continued parallel discussions of large-scale forest reserves and the more modest prospect of private forest tree planting.[35] Baker resumed his objection to planting trees on roadsides or railways because of the potential danger to travellers of accidents that could arise from the excessive downpours and falling trees. A new voice, that of John Crozier, another pastoralist, wanted to support the measure because of the shade produced for people, for stock and for other vegetation. He also wanted to revisit the evidence that trees caused rainfall, as he remained unpersuaded. In opposition to the botanical science presented by colleagues, he presented the work of highly respected Matthew F. Maury, a United States meteorologist who was instrumental in establishing an international collection of land and sea weather data using a set of agreed common standards.[36] Crozier cited Maury's southern hemispheric findings to support his view that geographical location on the globe's surface, rather than the presence or absence of trees, was the key determinant of a country's rainfall. Quoting extensively from Maury's publications comparing South America and the east coast of Australia, Crozier ended by refuting what he considered to be Baker's silly comparison of English and South Australian weather conditions, and supporting both the planting of reserves and the construction of roadsides.[37]

As requested, Schomburgk submitted his report to parliament on tree planting in October.[38] His "Report on Laying Out Forest Reserves, and Planting Trees on Railways and Ordinary Roads", asserted forcefully

---

[34] ibid.
[35] *SAPD,* "Planting of Forest Trees", 13 September 1870, 738.
[36] ibid.
[37] ibid.
[38] *SAPP*, "Report on Laying Out Forest Reserves, and Planting Trees on Railways and Ordinary Roads", 1870-71, no. 144.

that "the wasteful destruction of our forests" must stop, and "a scheme of both protection and production" must be instituted immediately. His primary arguments concerned climatic and economic transformation. He simply asserted the "recognized fact of the beneficial influence that trees have on climate, as well as their commercial value".[39] Schomburgk proceeded to recommend specific trees for specific locations. Each of the principle species named was native to Australia. Jarrah was the outstanding favourite, but also redgum for its "quick-growing" and "highly ornamental" features, tooart for durability, "fragrant sandal and raspberry wood", while the blackwood and the bluegum of Tasmania also came highly recommended for both function and beauty. Reflecting the current domestic and international practice of acclimatisation, Schomburgk went on to suggest European and North American trees for plains, hills and reserve planting.[40] These included the ash and elm, the locust tree, white cedar, poplars and pines. With additional resources, he said, the Botanic Garden could assume responsibility for experimenting with the cultivation and supply of up to 20,000 tree saplings annually. But, echoing Krichauff, he argued that the reserves would require the onsite expertise of men trained in forest culture to nurture the young plantations.

Goyder led the drafting of the Return to Order, the "Report on Suggested Forest Reserves" that had resulted from Krichauff's motion.[41] The Report was simply an extrapolation of his earlier report on the colony's wastelands.[42] However, in it, he also captured the intent of Hodgkiss' motion. He crystallised the significance of the issue of forestry but without any of Schomburgk's claims of its powers of climatic or economic transformation. The preamble succinctly restated the important elements of the government's resolution, trademarks of Goyder's professional communication. In it, he advised the "best size of reserves for forest purposes", where they would be, the "most economical means of preserving the native timber thereon, and of planting or replanting the reserves as permanent state forests". He also recommended both "the most valuable indigenous or foreign timber trees" for planting, aware of the need for both "a supply for public purposes as also annual revenue from the sale of surplus timber".[43]

---

[39] ibid.

[40] For more on acclimatisation see I. Tyrrell, "Acclimatisation and Environmental Renovation: Australian Perspectives on George Perkins Marsh", Environment and History, 10, 2 (2004): 153-167.

[41] *SAPP,* "Report on Suggested Forest Reserves", 1870-71, no.144.

[42] *SAPP*, 4 August 1870.

[43] ibid.

He also offered two new pieces of advice: that reserves should be no less than four square miles in size, and that they should be located in particular places in four disparate regions of the colony to provide for all settlers: northern districts, south-eastern districts, the Yorke Peninsula, and existing reserves for travelling stocks. The precision and detail of this operational advice continued to assume the necessity of establishing large scale afforestation practice for sustainable settlement as incontestable. It did not presume to advise on the contentious ideas raised in parliamentary debates about climate, but assumed the anticipated positive outcome of a vote for reserves.

Despite the broad-based support for tree planting, evident in the debates through 1870, and the confirmation of the wisdom of these measures in the concrete advice of the colony's leading public servant, no formal legislation was enacted that year. Members were too ready to engage in the elements of the debates about which they had opinions, with or without supporting evidence, and they were too ready to leave the big picture of forestry's role in sustainable settlement to the advocacy of Krichauff in parliament and Goyder and Schomburgk outside. The establishment and management of forestry as legitimate government business had some way to go.

Politicians seek local advice

The debate did not resume until the following year. In October 1871, Krichauff sought to contain the range of ideas available for debate by moving for a collection of evidence from district councils on the value to their region of promoting tree planting. He asked that a questionnaire, modelled on one developed by von Mueller in Victoria, be used to determine public opinion on the local economic value of forestry.[44] Concerned that "the culture of trees had not proceeded to any large extent" through the measures adopted last session, Krichauff hoped to inspire decision through action.[45]

With the motion declared, Krichauff then reprised the key issues that had arisen in the previous year's debates. Colonists would soon be at risk of losing their one reliable source of energy, he warned, because so many parts the public forests had been "annihilated" and there were no coal fields to act as alternative supplies of fuel. Forests did not belong to the current generation, but were "the people's patrimony". He reasserted

[44] *SAPD*, "Promoting Forest Culture and Rural Industries", 25 October 1871, 191.
[45] ibid.

his belief in the capacity of forest trees to alter the climate. He claimed that the collective weight of the international scientific evidence pointed to forests "ameliorating the climates". So, if only parliament would agree to plant the country 200 miles north of Port Augusta, "periodic floods and droughts" could be stopped. In a more direct tone than he had employed the year before, he asserted that trees would make the rainfall much more regular and forests could be "made to grow, notwithstanding the barrenness and bleakness of the land". To the transformative prospects of climate change and agricultural expansion he added the health benefits that could be expected from cultivating a "forest culture" throughout the colony. From Europe to the East Indies and New Zealand, he went on, the health-giving properties of trees had been demonstrable. The native eucalypts in particular had properties that tended to "arrest the miasmata floating in the air".[46] While Krichauff and Schomburgk were important champions of forestry, and their German scientific credentials tactically important to the cause, Goyder could not support all that they promised for forestry. He knew his own, much stronger empirical evidence that such promises were fanciful.

Having opened with an economic argument in the hope that it would capture even the most sceptical listener, Krichauff decided to conclude with another appeal to members' hip-pockets. If an earlier generation of colonists had had the forethought to grow forests, he lamented, then the current generation would have been spared significant public outlays such as those associated with the Burra railways, where costly iron posts had to be obtained from England in order to bring the ore mined in the district to port. As well as the prospect of greater self-sufficiencies of this kind, colonists could also reasonably expect flourishing industries in tea, mulberries and hops. Raucous shouts of "hear, hear" from members erupted when he spoke of a Gippsland farmer producing 1,500 pounds of hops on a single acre.[47] Breweries were an early and much loved industry in the Australian colonies, and one dependent upon sustainable supplies of hops. Forest reserves, broadly conceived, could begin the supply chain of these staples of colonists' domestic life.

Perhaps hoping to stem a rambling debate, the Commissioner of Crown Lands indicated enthusiastically at the conclusion of Krichauff's remarks that "officers of his department should think nothing of any trouble that might be involved" in sending out the circulars he proposed.[48] Another member suggested the wisdom of these same officers synthesising

---

[46] ibid.
[47] ibid.
[48] ibid.

and summarising the voluminous information likely to be collected before it was laid before the honourable members. The public, it seemed, would have their say on the matter of forestry and guide the decision-making of politicians.

However, before the discussion ended, one member dissented outright. The land agent and pastoralist James White thought it simply a waste of time in a tight labour market to try to build an export industry based on the need for afforesting the colony. In his view, trees should be cultivated merely for ornament, if at all, but as wool, wheat and copper would remain the chief staples for export, "timber-growing" would amount to time-wasting.[49] A challenge to Krichauff's logic and values rapidly ensued. Alexander Hay said he "could scarcely reconcile the honourable member's action now with his support of a provision regarding clearing and grubbing as an improvement". He queried how could he advocate tree planting on the one hand and tree removal on the other. Arthur Lindsay leapt to Krichauff's support, saying he, too, could be accused of the same charge of inconsistency. But, he retorted, "trees which would be very valuable in particular localities would be an encumbrance in others".[50] The radical land reformer Ebenezer Ward, noted for his eloquence, also reminded Hay that, in their considerations of this and other matters, they must be more complex in their thinking than his objection suggested was the case. It was, he pronounced, "in discrimination that they showed higher intelligence, and in clearing land that it was desirable to clear, and cultivating forest trees where it was desirable to cultivate them, they showed wisdom".[51]

Krichauff countered the challenge with less profundity. He believed simply that a "tree in its place was a very good thing, but a tree in a corn-field was not in its place".[52] This appeal to members' less earnest sides resulted in another round of cheers. As to the supposed dearth of labour available for tree planting, Krichauff saw instead the possibilities of employing farmers during fallow weeks, and the rehabilitation through community engagement of orphans and reformatory boys.[53] With this comment, the matter was put aside, while the district councils, with the support of central government, went in search of public opinion on forestry. Krichauff and Goyder now waited on the outcome of the strategy

[49] ibid.
[50] ibid.
[51] ibid.
[52] ibid.
[53] ibid.

to garner a broad public consensus in the hope persuade parliamentarians to act on the innovation.

The government must act alone

Six months later, Krichauff again rose in parliament as forestry's vocal champion. Patience and perseverance were to prove essential in the armoury of the innovation of forestry for both Krichauff and Goyder. But Krichauff's patience was beginning to wear thin. He opened with the issue of governance, local versus state, and expressed his frustration with the general inertia in the matter of tree planting and protection.[54] Having had little success in engaging the district councils with his questionnaire, he now proposed that reserves should be made the responsibility of central government. Krichauff argued that councils, far from being supportive of forests in South Australia, were unwilling to prevent residents destroying trees and de-pasturing stock on forested areas of public land.[55]

This new motion presented a direct challenge to the vested interests of the colony's landed aristocracy represented by the pastoralists in government.[56] The fundamental tenets of individual land ownership and land rights were contained in Krichauff's proposal. As a successful pastoral lessee in the mid-north, William Mortlock stridently opposed the challenge to his autonomy to use of the land and its resources in whatever way he chose. He claimed that "the Northern areas would be useless if the settlers were not allowed to cut down the timber to fence their lands".[57] It was imperative that the proposed reserves be entirely open for settlers to use, especially in the north, where there was simply no timber in many parts. He declared that the government "might as well burn the Agricultural Areas Act for all of the use it would be" in protecting his interests.[58] But, in more measured tones, John Angas, another member of the South Australian landed gentry, sought to bridge the polarised positions of unconditional access and no access represented by Mortlock and Krichauff. He noted the benefit to settlers, and the lack of damage to forests, of allowing settlers conditional access to poorer timber in the forests. In fact, he said, it was vital to the survival of settlers in many areas

---

[54] *SAPD*, "Forest Reserves", 3 May 1872, 773.
[55] ibid.
[56] J. Hirst, *Adelaide and the Country 1870 – 1917* (Melbourne: Melbourne University Press, 1973) 227.
[57] *SAPD*, 3 May 1872.
[58] ibid.

where the country had been "denuded of trees" that such conditional exclusion be implemented at once.[59]

Months later, and close to two years after their initial motions in their respective houses, Hodgkiss and Krichauff rose yet again to move for public funds to realise the measures they had first championed. Hodgkiss asked for money to plant trees along railway lines;[60] Krichauff wanted 500 pounds for the general purpose of "protecting and planting forest trees".[61] Krichauff appealed in his motion for the direct patronage of "His Excellency the Governor to recommend this House to make provision in any Supplementary Estimates".[62] His lack of faith in the House to deliver even a modest amount for planting was matched by his frustration with Schomburgk for his tardy delivery of yet another report. Krichauff insisted on not losing another season's opportunity for planting and pointed to the activity in both New Zealand and California, which he felt should move South Australians beyond their present comparative inertia.[63]

The Commissioner of Crown Lands denied his request, citing the fact that Schomburgk had yet to deliver his report. The tensions between Krichauff's passion to realise a German forest culture in his adopted home and the realities of competing government priorities were evident, even in the midst of a largely supportive administration and parliament. In the absence of a critical scarcity of timber, the combination of Krichauff's, Goyder's and Schomburgk's reasoned approach had been unable to generate an energetic vision for action on forestry to match their own.

[59] ibid.
[60] *SAPD*,"Forest Trees", 16 July 1872, 1301.
[61] ibid.
[62] ibid.
[63] ibid.

CHAPTER 5

DESPERATE TIMES LEAD TO MODEST MEASURES

The tension between the general mood of support for the innovation of forestry and the inability to get legislation passed to secure it produced an increasing tone of desperation in Krichauff's parliamentary speeches. He thought the procrastination shown by the government in the leadership of this critical innovation was no longer tolerable. In a move of apparent desperation for decisive action, he introduced a new bill in 1873 that gave away the prospect of large-scale public forestry. By December 1873, Krichauff had succeeded in getting legislation through parliament that provided a direct incentive for individual landholders to dedicate significant areas of their land to forest trees. Clause 4 of *An Act to encourage the Planting of Forest Trees* summarised its essence:

> if any person shall plant any land, not being waste lands of the Crown, and not being less than five acres in extent, with forest trees, he shall be entitled to receive, in respect to every acre so planted, a land order in the form in the Schedule hereto, which shall authorize such person to purchase, to the amount of Two Pounds, any of the waste lands of the Crown.[1]

But even this modest outcome required reprising much of the substance of earlier parliamentary debates. An overview of this renewed effort on behalf of forestry follows. Once again, the messiness of competing ideas is mapped as further evidence of how an environmental wicked problem was played out in the houses of parliament.

Krichauff and Goyder renew efforts

Introducing the new bill, Krichauff returned to his favoured international comparisons, this time incorporating more references from the United

[1] *An Act to Encourage the Planting of Forest Trees,* no. 26, 1873.

States of America. His focus was forestry-induced climate change. He first cited Napoleon III, who had planted the vast forests of Algeria with eucalypts (to the House chants of hear, hear from members) with the effect that the quantity of rainfall was twice what it had been before the forests. Were Napoleon's example to be followed in the colony's northern districts, advised Krichauff, it seemed very likely that the "extreme changes of climate which they so often witnessed… would be no longer endured".[2] He drew again on earlier comparisons from New Zealand and the East Indies, which demonstrated the direct connections between tree felling and a drier climate.

Perhaps because of the parallels to be drawn between the former British colonies of the United States and those of Australia, evidence from the United States of America increasingly appeared in debates. Krichauff provided information from the relatively recent international data from 1868 reports by the Land Commissioners of the United States. These noted that in several of the "Far West" states, floods had "very greatly diminished since settlement had set in and the treeless prairies had been largely planted".[3] The Commissioners had also calculated that, if the present rate of tree felling had been permitted to continue, supplies would be exhausted by 1888. Iowa shone as an example of remediating this disastrous prospect by planting seventy-six square miles to trees.

The combination of the scientific training Krichauff had experienced and the "bird's eye-view" method followed by Goyder in the determination of the colony's social planning needs were beginning to making some progress for forestry in South Australia. A week after Krichauff's first reading of the new bill, Goyder submitted a further report upon the same subject to parliament.[4] In doing so, he was able to lend the weight of empirical evidence of the physical environment to the international examples favoured by Krichauff. He had also begun to use impassioned language, noting that the "rapid decrease in forest trees is brought painfully and prominently before me" in the North. But this was now true in "whatever direction my duty takes me". Parliamentary debate had concentrated its efforts regarding tree planting in the North, the obvious direction in the colony for immediate agricultural expansion, but Goyder took this opportunity to renew his earlier efforts on behalf of other areas which were not so obviously affected. He observed that "in the South-East District the whole of the indigenous timber is rapidly dying".

---

[2] *SAPD*, "Planting of Forest Trees", 10 September 1873, 272.
[3] ibid.
[4] *SAPP*, "Report on Forest Trees Planting Encouragement Bill" no. 94, 18 September 1873.

This was occurring "to such an alarming extent" he feared that "unless remedial measures are adopted, a greater portion of same will be treeless". Healthy timber, he continued, did exist but was the exception in the colony. It was imperative that all three steps of "planting, protection and judicious management" be immediately employed. Goyder noted the "universal decay of large forest trees, especially of the various kinds of gums in this Province and Victoria". He considered that neither drought, an excess of moisture, damage done by animals, such as opossums, nor the relative age of trees played a part in the widespread decay. He hypothesised instead that the cause was most probably the exhaustion of certain ingredients in the soil essential to the growth of eucalypti, and recommended an analysis of soils be undertaken to ensure the profitable match between soil and tree and/or the necessity of supplying the exhausted substances through "properly constituted manures".[5]

Goyder commended Krichauff's bill as a "step in the right direction". In dealing with planting on private property it would go some way to rectifying the deleterious practices that had led to the large-scale damage he had witnessed. But it would be insufficient to depend on private land owners, interested in maximising their market yield. The most important subject before the government, he declared, was the desirability, if not yet the absolute necessity, of proclaiming and sustaining forest reserves on Crown lands. Land must not only be resumed and protected from "the indiscriminate depasturing of sheep and cattle" but must also have inspectors attached to it to monitor the growth and development of the young trees determined as suitable for the particular locality. Schomburgk's idea of gardeners cultivating seedlings in flower pots for the annual distribution of twenty- to thirty-thousand saplings throughout the province may be adequate for road or railway planting, but could not, declared Goyder, address the extent and diversity of the need. Instead, he argued, young plants should be raised in large numbers in dedicated nurseries established close to the proposed plantations, and these nurseries needed specialist, local knowledge to man. They must, he said, "be fixed by men accustomed to the cultivation and growth of forest trees in all kinds of country and classes of soil".[6]

Concluding this part of the report, Goyder praised the work of just such a local man, Mr Edwin Smith, "a nurseryman of large experience" (and his brother-in-law), whose recommendations he intended to adopt and formalise in regulations to accompany his final report.[7] It

[5] ibid.
[6] ibid.
[7] ibid.

seems probable that Goyder saw Smith's local knowledge and experience as desirable in achieving the large-scale forest management he hoped would follow in the colony. He may have had him in mind for the position of the colony's first Conservator.

Krichauff's impatience with Goyder's patience

But Krichauff seemed unpersuaded by the measured, domestic focus of Goyder's report, however urgently it pressed for reform. The scientific theory of the causal relationship between trees and rainfall was, for him, indisputable in the multiple international examples he had collected and offered to his colleagues. Also, it offered a much grander argument for forests than Goyder's. With growing impatience about the prospects of large-scale forest reserves, Krichauff shifted his entire focus to the issue of encouraging private citizens to plant trees, proposing a scheme to bring this into immediate effect. The Chief Secretary asked him to postpone his motion.[8] It is possible that Krichauff's apparent frustration could have been a tactical ploy, designed to keep the matter of forest tree planting alive before an ambivalent parliament as Goyder travelled extensively across the colony to supply the latest local environmental data to parliament.

However, whether tactical or heartfelt, Krichauff's impatience once again gave rise to lengthy debate.[9] The lawyer James Boucaut was persuaded of the certainty with which Krichauff presented his arguments for transformative climate change. He wanted him to proceed directly, without waiting for Goyder's further report, to lay down a motion to "set aside forest reserves as was done in Canada and other thickly wooded countries". Krichauff instead wanted to return the discussion to measures that would affect private property. He pointed to the success of New Zealand's Acts of 1871 and 1872, which had resulted in large numbers of farmers planting trees for the remuneration of three pounds per acre. But Angas also wanted to take a holistic view and the potentially catastrophic consequences of failing to act on a large scale. He warned that "unless land was enclosed for arboriculture the trees would die out, and the country be materially altered".[10] His own travels in Egypt offered evidence of the increase in rainfall experienced in that country by the Pasha's planting. Smith then tried to thank Krichauff for his perseverance and suggested that "a difference had come over the temper of the House since

---

[8] *SAPD*, "Encouragement of Tree Planting", 24 September 1873.
[9] ibid.
[10] ibid.

last session".[11] This was demonstrable in the sums approved for Adelaide's, and other councils', local tree-planting. But Angas became impatient with the inconsequential nature of such modest change. The scope of the issue, he protested, was much larger than planting a few trees along roads in townships and dealing with the "mischievous boys [who] nipped the tops off" only to be fined the small sum of one shilling.[12] Whether tactical or not, this pendulum-swinging debate between small-scale private forest tree planting and much larger public innovation resulted in a committee charged with drafting and bringing in a Bill that would promote the private business of tree planting, while advancing measures to resolve the debate on forest reserves.[13] The political machinations to effect the intentions of Goyder's report on forest reserves of 1870 were moving ahead as political awareness evolved. And, whatever their differences of scientific opinion about forests and their climatic benefits, it was the shared advocacy of Krichauff and Goyder which mattered at this stage of innovation.

Goyder's final report

By the end of October Goyder had a final report ready for Parliament. This was submitted for consideration on 12 November.[14] Having witnessed the uncertainties of debates of the previous three years it was now time to provide an elaborated version of his understated bill of 1870 Land Bill, where he had said simply that "these reserves, as they are denuded of timber by the occupiers of the open grounds, should be again systematically planted – as is done in many places".

With typical succinctness, the report presented fifteen clear recommendations.[15] Again, it named the location of the reserves in northern, southern and central locations and contained the new detail of exact placements over a total area of three hundred square miles. Once formally declared under the Waste Lands Alienation Act of 1872, regulations for their proper management were to be compiled and ratified by the Governor in Council. Eight of the recommendations declared what Goyder considered to be foundational to successful management. An officer, "thoroughly and practically acquainted with forest culture in all its branches", must be appointed as Conservator of Forests. He would become

---

[11] ibid.
[12] ibid.
[13] ibid.
[14] *SAPP*, "Report on Forest Reserves" no.135, 1873, 12 November 1873.
[15] ibid.

the advisor to the Commissioner on all such matters, be instructed to prepare lists of seeds for ordering, obtain the necessary labour to establish nurseries, beginning with three, one in each region and building gradually to seven once sound propagation procedures had been determined for the distinct environments. Two principles would govern the selection of seeds and the establishment of nurseries. Each assumed a preceding principle that forest reserves were part of a well-functioning system of human settlement, not an end in themselves, as Goyder made clear in recommendation 5.

It simply made no sense to Goyder to quarantine human activity from forestry activity in the settling colony. In an 1872 motion, Krichauff had sought to separate human and natural environments by excluding forest reserves from the survey of land for settlement purposes.[16] But Goyder courteously pointed to the supersession of this agreed resolution of the House of Assembly by the Waste Lands Alienation Act of 1872, which decreed that all crown land was available for government survey. He noted that timber would be required by those who settled in the hundreds and that its proximity would be convenient for them to access it simply made sense for forest reserves to be included in the hundreds his department would continue to survey for occupation.[17] There was no doubt in Goyder's mind that the forests must be put to work in the service of settlement and that species' selection would prioritise what would grow best, fastest and to most advantage. Ornamentation would be of secondary importance. The first season would see one million seeds planted to an initial protected area of seven thousand acres, seeds and acres increasing in annual increments. Demonstrating his own and Edwin Smith's expertise in the field, Goyder went on to stipulate even such detail as the number of trees to be planted per acre (435), plantings at every ten feet apart, and thinning and sale as specified by the Conservator in five yearly cycles until the distance between, and quality of trees, allowed for the growth of a mature forest. In all of this activity, the Conservator was to exercise precision in planning and execution.[18]

The funding of the reserves, the final and most difficult hurdle of persuasion, Goyder left to last. He approached this sensitive element of the argument by providing the bad news on projected expenditure first, and the good news of prospective revenue second. Establishment costs for the first year would necessarily be higher than in subsequent years, he declared, and would mostly consist of wages and accommodation for the

[16] *SAPD*, 3 May 1872.
[17] *SAPP*, 12 November 1873.
[18] ibid.

required numbers of men and horses. One-off costs for implements, fencing and flower pots must also be taken into account. Goyder's carefully itemised budget tallied to an initial expenditure of 14,357 pounds and 10 shillings, reducing to 10,500 pounds in the second and subsequent years. But against this he projected an income after five years of 35,000 pounds from the sale of 3.5 million trees, rising to 70,000 pounds by years 10 to 12. This was an ambitious claim, unlikely to come true, but Goyder used it to make the unequivocal statement of the profound significance he attached to the innovation of forestry for a sustainable society.[19]

Finally, it was left for him to clarify the distinctive purposes of his report and of Krichauff's Bill for tree planting on private lands. Whether the final distinction to which Goyder alluded was tactically manufactured, or the combination of differentiated effort, proved effective in sustaining parliament's increasing willingness to act on forestry – but it was not going to do so without further debate.

Goyder's recommendations: "Report on Forest Reserves", 1873

Number	Recommendation
1	That the reserves be divided into three classes-northern, southern, and central-as follows, -Northern: Never Ranges & Springs, Hundred of Bundaleer; Wirrabara Forest, south of White Park, east slopes of Mt Remarkable, travelling stock reserves in Hundreds of Wongyarra, Appila etc (total 27) -Southern: Mts Gambier, Burr, McIntyre, Muirhead Flat, south of Rivoli Bay, Cave Range, Cockatoo lake land, Glen Roy Flat, Border Town, travelling stock reserves X 4 -Central: Govt Farm Reserve, Waterfall Gully; Mt Lofty Reserve; travelling stock reserves X 15 Total area proposed 300 sq. miles.
2	That the Reserves in question be proclaimed in the *Govt Gazette*, as provided in Section 9, Waste Lands Alienation Act, 1872-and that the regulations for their management be compiled in terms of clause 51-which enables the Governor in Council to make and publish regulations for carrying out the object of the Act (N0.18 of 1872): Provided that such are laid before Parliament.
3	That an officer, thoroughly and practically acquainted with forest culture in all its branches, be appointed as Conservator of Forests; and that he be instructed to prepare lists of seeds to be ordered, and to obtain the necessary labor to establish nurseries for forest trees at Mount Gambier, in the Southern District; the Springs, Bundaleer, in

[19] ibid.

	the Northern District; and at the Government Farm, in the Central District-so that a commencement may be made without delay, and that the nurseries be gradually extended to the Cave Range and Border Town in the South, and Wirrabara Forest and Penwortham Reserve in the North-and making a total of seven nurseries to supply stock for the three districts, *i.e.*, three for the South, three for the North, and one at Government Farm. For the Central District.
4	That the duties of the Conservator of Forests be to advise the Commissioner on all subjects connected with forest culture.
5	That the formation of forests be proceeded with gradually, and in such a manner that, whilst a thoroughly organized system of raising, planting, and protecting be at once provided for, care be taken not to unnecessarily interfere with free access to existing forests; or, to render it more difficult to persons to obtain and remove timber than can be done by the public at large, under the present regulations-for instance in the Wirrabara Forest such portions only will be first planted and protected from which the valuable timber has already been removed, and where only sufficient remains to form protection for young stock. A similar course being adopted in all cases where existing forests are now being made use of.
6	That the trees selected for cultivation shall, in all cases, be such as are best adapted for use in manufacture, building, fencing, mining, &c., and not merely for ornamental purposes, and that they comprise various eucalypti, such as jarrah, red and blue gums, tooart, &c.; pines, such as Pinus insignis, maritime, canariensis, halepensis, and Waymouth; with oak, beech, walnut, sycamore, poplar, willow, cedar, chestnut, ash, sandalwood, olive, and locust trees. As the sheoak and blackwood will spring up abundantly where the land is protected from stock, it is not considered necessary to include them in the list.
7	That, in the reserves, trees be planted in the first instance ten feet apart, or say 435 to the acre. That in five years time every second tree be removed and sold for rails, or other suitable purpose, leaving the remaining trees twenty feet apart. At the end of another five years a similar thinning out and sale to take place, leaving the remaining trees thirty feet apart, when that portion of the forest may be considered established, and the trees be allowed to attain maturity, and only to be removed on being properly marked for falling by the Conservator.
8	That sufficient area be at once protected, viz., 7,000acres, or say eleven square miles, to receive the plants that would be raised during the first season at the three nurseries…say 1,000,000 plants from each, and that such area form a portion of each of the reserves in the respective districts, and that the area protected be increased each year, so as to be capable of receiving the stock raised from the nurseries as they are increased, until the whole seven have been established.

9	That the Conservator be required to keep records of the several portions of each forest as they are planted, and of the trees planted therein, so that no doubt may exist as to the ages of trees grown, and that the requisite sequence be preserved, and the different portions of the forest availed of in proper rotation. These records should be illustrated by a plan, showing the portions of reserve planted and protected, and the portions from which timber of various kinds may be removed, and under what conditions.
10	That, as most of the travelling stock reserves comprise rich lands upon open plains, and upon which, from their bleak nature, it might be difficult to grow young trees without protection, the Conservator should be required in such localities to plant suitable shrubs of rapid growth, which might be permitted to remain until the young trees attain sufficient strength to do without such protection.
11	That prior to the fencing in of any forest reserve, roads be laid out so as to give access to each portion of such reserve, and that the fences be so placed as to obstruct the traffic as little as possible-proper swing gates being placed upon such roads: this is the more essential in reserves for travelling stock, as in many cases the best lines of road wind from one side of the reserve to the other, and rarely continue in the middle or on one side.
12	The present system for granting licences for cutting and removing timber from Crown lands must remain intact, until the timber upon the protected portions is ready to be cut, when Regulations will require to be framed to meet all the circumstances of the case; meanwhile it will only be necessary to exempt by strict Regulation and *Gazette* notice, with heavy penalty for infringement, those portions of the reserves that are from time to time planted and protected.
13	That forest reserves be allowable within Hundreds, though excluded by the H0A on Mr K's motion (of 3.5.72). The resolution in question was passed prior to Waste Lands Alienation Act, 18 of 1872, which includes all land within the First Schedule of that Act, as liable to be surveyed for selection under its provisions. This, to a certain extent, makes it imperative that the whole of the land within the limits of that Schedule shall ultimately be proclaimed into Hundreds, otherwise the land could not be open for selection; to exempt therefore forest reserves from Hundreds, since the passing of the Act, would be either to place such reserves outside the Schedule, or require the revision of a large number of Hundred boundaries…and, inasmuch as timber will be largely required by persons holding lands within the Hundreds…and will be doubly convenient and valuable, from their proximity to agricultural lands almost entirely destitute of timber.

14	Assuming that the above suggestion be adopted, and a Conservator of Forests, with the requisite staff, for the seven nurseries appointed, the probably expense of the Department for the first year would be 14,357 pounds 10s. (cons. 400, 7 gardeners for each nursery, one labourer, 2 boys, 2 horses and cart (for each), shed & hut (for each), nursery implements, seed, travelling for Con, flower pots, fencing, extra assistance during planting season) - [expense for second and following years expected to be approx. 10,500 pounds] - This amount need not be at once expended if the suggestion of commencing with but three nurseries be adopted.
15	The seed which cannot be obtained in the Province should at once be ordered from England.
16	Five years must elapse after commencement of operations before any return on the outlay can be expected; then, assuming that the whole seven nurseries be established at once, and the 7,000,000 trees be planted during the first planting season…every second tree…might be removed and sold at 1 pound per hundred, realizing 35,000 pounds in cash. This amount would continue to be realized for each of the following years, when it would be largely increased by the removal of every other tree from the block first planted, and which would then remain a forest- the trees being forty feet apart, and only cut on arriving at maturity, and as marked for sale by the Conservator. The expenditure to carry out this scheme, therefore, would, during the first year, be, say, 14,500 pounds, the eleven following years at, say, 10,500 pounds a year (115,500 pounds). Total expenditure in twelve years, when all the forests would be fenced and planted, 130,000pounds. The revenue derived during the same period would be for the first 5 years 0, 6-9th 35,000 pa., 10-12th 70,000p.a. 70,000pounds would continue to be the revenue until the end of the seventeenth year, when it would again decrease to 35,000pounds, that is, to the end of the twenty-first year, when the country would be in possession of three hundred square miles of matured forests. Of course, the above calculations do not take into account failure here and there in the growth of trees, but they are sufficient to show the value of the work, and that there is every probability of its being self-supporting in addition to its vast importance. As this scheme in no way interferes with Mr Krichauff's Bill, it is not alluded to in the present report.

Krichauff promotes private forestry

Krichauff's opening remarks a week later stressed the moral imperative of private citizens planting trees.[20] The Government might be able to "plant some forests", he said, but "private persons must be looked to for something".[21] Choosing as his focus the imminent exhaustion of timber for fuel supplies and the financial stupidity of looking to importation as an option, Krichauff began by challenging Goyder's optimistic income projections. He refuted these based entirely on examples of slow growth European species. In his objections he completely, and oddly, omitted any reference to indigenous trees. This was perhaps an indication of a genuine conflict with Goyder, in which he sought to replicate his German experience of forestry locally. He went on to promote tree planting around the perimeter of fields, as had occurred in several states in America. If Americans, he asked, with their 1,400,000 square miles of treeless plains, had decided to do this, how much more important was it in this colony where the climate was so dry?[22] Advice he received from the USA had also indicated that the clear felling of forests in Pennsylvania, Ohio and New York had dramatically altered the climates of these states from what they had been fifty years earlier. In this, there is no doubt that his views and those of Goyder were directly opposed. Returning to the immediate and pressing problem of the want of basic fuel, Krichauff declared the foolhardiness of expecting timber imports to remedy local need. Apart from the unnecessary waste of money, these also looked set to dry up. He quoted directly from an article in the *Australasia*, to substantiate his warning:

> There is a mania just now for clearing forests, coequal and coextensive with the desire for possession of the land. The question of timber supply belongs not merely to one or two countries, but to the world at large. The destruction of forests appears to accompany what is called European civilization. In Japan, as in this country, the timber is being fast cleared off in order to afford space for crops, which yield an annual return for the outlay. Many decades will have to elapse ere the mischief done can be even partially repaired.[23]

---

[20] *SAPD*, 3 December 1873, 1314.
[21] ibid.
[22] ibid.
[23] ibid.

Krichauff noted authoritative advice that if the present rate of consumption continued in the USA it was predicted that “the whole of the forest east of the Mississippi would be cleared in the early part of the next century”.[24] With Norway and Sweden also reporting dwindling timber supplies, Krichauff hoped that the colony would, “like other enlightened nations, take up the question and resolve to act”. Seemingly taking whatever opportunity was available to do so, he declared his belief that only the planting of trees would “cause the rain supply to be large”.[25] Such views were not uncommon amongst forestry’s champions around the globe.[26]

Commentary following Krichauff's remarks was both broadly supportive and deeply nit-picking. The timber merchant James Pearce went straight to suggesting a preference for the gum of Tasmania because of its rapid growing properties. Ebenezer Ward wanted to dismiss the bill entirely as one which would result in the mere planting of shrubberies and replace it with one for State forests. The lawyer and future Premier John Bray suggested Ward calm down as “Goyder’s scheme” was not interfered with by Krichauff's bill. He went on to say that the regime proposed in the latter was open to corruption and exploitation. The Chief Secretary concluded by commending the continued commitment of both Krichauff and Goyder to the cause of forestry and was “astonished only that the work had been so long left undone. The bill was passed on the second reading.[27] However, the bringing in of the Bill could not be taken for granted just yet.

Krichauff was proposing to legislate, as New Zealand had done, for the payment of four pounds per acre to any farmer planting forest trees on his property, leased or owned. Members wanted to contest the minimum number of acres to be planted, the minimum number of trees to be planted per acre, the definition of what “forest trees” were, the amount to be paid, when said amount was to be paid, whether inputs (saplings) or outputs (grown trees – even of specific height and girth) counted, the nature of nefarious activities in which farmers could engage to subvert the system, and other related matters.[28] Amidst this contest of detail, Krichauff introduced a red herring, perhaps to try to return some perspective to the debate. He wanted it known that the East Indies had planted two and a half million cinchona plants, and hoped the government would extend the

---

[24] ibid.

[25] ibid.

[26] C. Miller, “Amateur Hour: Nathaniel H. Egleston and Professional Forestry in Post-Civil War America”, Forest History Today 25 (2005): 20-26.

[27] *SAPD*, 3 December 1873.

[28] ibid.

provisions of this Bill to encourage similar large scale planting in its Northern Territory. Mortlock threw in his own red herring. With deep irony, he expressed concern that pastoralists, such as himself, would be exempt from the remuneration offered, as their lease would likely have expired before payment was made, given how long it would take them to grow any tree to maturity. He also declared with more serious irritation, the "squatter" did not want compensation "and the House were taking a great deal of trouble for nothing".[29] Individual clauses of the Bill continued to be argued in earnest for days, and at length, while Mortlock continued to lampoon proceedings. In discussing the quality of the tree for which remuneration might be paid, Mortlock joked that height was a fallible measurement. He had some trees he estimated to be ten feet tall, or at least twice the height of the member for the Barossa, but he thought their vigour may be as questionable as the honourable member's. In final dismissal of the fundamental premise of the Bill, he declared with almost palpable exasperation, "if the pastoral lessees wanted to grow trees let them, but it was no use trying to grow ten-foot trees where nothing but scrub would live".[30]

But the perseverance of forestry's advocates finally won out so that by the end of 1873, the colony had an Act to encourage the planting of trees which would ironically, in the long-term, prove Mortlock right (*An Act to encourage the Planting of Forest Trees, 1873)*. Only two applications were ever received for the agreed remuneration of two pounds per acre, and only one of those proved successful. Private landholders could not be persuaded of the environmental or economic basis for altering their focus on crop growing, even though one of the clauses specifically allowed for the co-existence of forest trees and root crops on the same patch of land. Clause 4 stipulated that

> provided also that if root crops are cultivated amongst the forest trees planted, such land shall not, by reason only of such crops being cultivated, be deemed to be not devoted only to purposes of planting.[31]

While this modest Act did not see the flourishing of forestry in the colony, it did nurture "Goyder's scheme" and the promise of something much more significant in the future. The past three and a half years had indicated that the protraction, perseverance, repetition, and elaboration of ideas and

[29] ibid.
[30] ibid.
[31] *SAPD* "Forest Bill" 17 December 1873, 1307.

debates was the necessary price to be paid for one Act of doubtful efficacy for private tree planting, and a parallel evolving awareness and acceptance of tree culture. The latter had enabled Goyder's incremental development of the operation of public forestry. There had been little dissension from the view that forest tree planting was a useful activity that would benefit the colony in one or more ways. But the harder edges of members' vested interests, or simple stubborn commitments, remained a constant backdrop and obstacle to the sort of progress Krichauff wanted.

There was still much to be done to implement public forestry on a large scale in the colony. The next stage would prove to be just as protracted as cultivating the idea of innovation in parliament. If the debates from 1870 to 1873 had managed a modest, symbolic Act and witnessed a ripening of the idea of forestry, the debates about the leadership of its implementation in 1874 and 1875 were to prove just as messy, circuitous and liable to generate inaction not action.

CHAPTER 6

FROM ENCOURAGEMENT TO MANAGEMENT OF FORESTS

This section concludes the account of the first phase of the innovation of forestry in Australia: securing its early legitimacy as public business through legislation. The often synthesised coverage of the parliamentary process disguises the realities of leading the complexities of innovating new ideas towards agreement to act. The chronicling of this often tortuous journey is intended to indicate how much the leadership of the innovation owed to the sheer perseverance and commitment to democratic processes necessary to achieve it. It further indicates the willingness of those charged with the stewardship of the public good in both houses of parliament to lapse into the contest over detail and avoid the apparently harder task of holding on to the larger vision to locate common ground. The commitment to an adversarial process of debate is, itself, a significant part of the wickedness of such problem solving.

The recurrent debates from 1870 to 1873 had resulted in modest legislation. By securing a reward for private landowners who planted a percentage of their land to forest trees, the Act had also implicitly declared forestry to be a significant element of settlement. The legislation nurtured the evolving parliamentary awareness of the need for large-scale forestry on public lands. A year after the introduction of the first Act, the issue of public forestry was returned to the parliament to try to secure its operation at a larger scale. Once again, Krichauff championed a bill drafted by Goyder and his staff. In July 1874, he introduced a Forest Board Bill that aimed to galvanise the idea of forestry through a system of governance and implementation.[1]

Unlike his previous introductions, which typically opened with dramatic international examples designed to motivate colleagues to action, this time, Krichauff first focussed on the local and pragmatic. He appealed directly to economic efficiencies. His principal reason for the bill, he said,

[1] *Register*, "Forest Management", 14 July 1874.

was to prevent the waste of the colony's funds. For the past two years at least, he noted, "considerable amounts had been voted in connection with forests, and he feared unless some persons were appointed to expend it" efforts would fail from a lack of proper management.[2] He declared that staff in the Commissioner of Crown Lands Office, under Goyder's supervision, were already greatly extended with no capacity to absorb the vastness of this critical work.

If Krichauff had expected to expedite debate through this practical approach, he failed. The debates of recent years had shown members in both houses to have a general sympathy with the big environmental and social ideas of forestry. It was precisely when debate turned to the operational detail that the proliferation of competing views emerged, leading to paralysing inaction. The convoluted and often petty contests so characteristic of the last three years were once again on display. The misjudgement that Goyder may have made is to expect that the passage of the bill would be expedited by providing a highly detailed draft, containing multiple clauses, that spelled out the governance and implementation of forestry. His assumption overestimated the readiness of members to engage again in the fine detail rather than the larger purposes. It took fourteen months for the bill to pass through the lower house, and even with the comprehensive coverage of issues that such a time span indicated, the upper house was not prepared to rubber stamp it.

Beginning at the end: the upper house speaks

Sir Henry Ayers, Chief Secretary and sponsor of the bill in the Upper House, used his substantial reputation as an esteemed senior colonist to support the measure. Ayers prefaced its second reading in eloquent, if rather condescending, tones. The passing of the Forest Trees Act of 1873 had been all well and good as far as it went, he said, but there had been no "machinery for carrying it into effect".[3] Rehearsing again the arguments for forest tree culture, he considered them redundant. He did not "suppose that there was any need for him to advance arguments in favour of the planting of forest trees". He acknowledged the doubt that existed about forests' capacity to alter rainfall, but assumed there was no doubt "in the minds of honourable members as to the advantage of planting forest trees". He declared that both their beauty and their usefulness were self-evident.[4]

---

[2] *SAPD*, "The Forest Board Bill", 1 July 1874, 784.
[3] *SAPD*, "The Forest Board Bill", 7 September 1875, 922.
[4] ibid.

Ayers referred tactically to examples of extensive tree planting by esteemed colonists, and highlighted the Krichauff's authorship of the bill, rather than Goyder, who was not well liked by the wealthy pastoralists in the House.[5] He referred members to Krichauff's presentation to the Chamber of Manufacturers, should they need any further evidence in support of the present bill. Here was their propertied friend in the Lower House offering sound economic reasons for advancing forestry. Surely members could trust his credentials.[6]

But, if he had been hoping for a token debate before rubber stamping, he was disappointed. In both Houses, newly elected representatives wanted to have their say, even if these same points had been made. The still-novel role of forestry and its science rendered Ayers' aspirations for brevity hopeless. Members voiced their ideas, their ideologies, and their brand of democracy in both cool and heated tones, and at length.

Thomas Hogarth, pastoralist, road builder, miller and machine-maker, led the charge against it on behalf of the practical working man. Not for him this "everlasting talk about science without practice". The so-called scientific evidence was perfectly obvious to them all. Everyone knew that when old trees were cut down others would grow in their place; that if they planted trees on the plains grass had to be kept down or they would burn; and that sheep given half a chance would eat the trees. They did not need the pronouncements of experts to make these facts true. Added to this practical knowledge was that of all farming men: that it was "ploughing that caused the rain to sink into the earth, and then must find its way out in the shape of vapour". Scientific expertise and its dispensation through government boards was nonsense, he said, an invention of the present age. In fact, "it was the practical man who had done everything for South Australia, not these shams".[7]

But pastoralist William Everard was to have none of Ayers' paternalism or Hogarth's earthy smugness. Ayers should have summarised the arguments in favour of the bill, he said, especially relating to the effect of forests on climate. The international evidence, he believed, was indisputable. Furthermore, Hogarth's championing of the common-sense land management of practical men had been very lopsided indeed. "The practical man whom his friend Mr Hogarth was so eloquent about", ventured Everard, "had done a great deal in cutting down trees, and he failed to tell the Council how much wood was wasted for every 100 rails

[5] Sheldrick, *Nature's Line*, 127.
[6] *SAPD*, "The Forest Board Bill", 7 September 1875.
[7] ibid.

obtained".[8] Decisions about forests deserved the earnest consideration of a disinterested citizenry and he "was not one of those who thought they should plough up everything, and do nothing but raise so many bushels of wheat".[9] To do so would be deserving of the scorn that subsequent generations would no doubt heap upon them.

John Crozier, a faithful advocate of tree culture, introduced drama to debate by referring to the *Sydney Mail*'s publication of an article from the international paper, the *Iron*. The writer argued that only in South America were trees present in excess, but that this was of little value because of the inaccessibility of these "primeval forests" to export markets. Norway and Sweden, principal sources of the pine essential for ship and house building, were now prohibiting the felling of trees below a certain girth, which was also likely to raise the cost of pit timber, and subsequently coal. Norway had already voted for public funds, established arboriculture schools and nurseries, and appointed officials to manage reforesting. Austria, Germany and France were also earnestly engaged in similar pursuits and, he reminded members, the citizenry of the United States had been clearly shown to be "alive to the dangers that menace them in this direction, and are proceeding too with judicious measures of protection and replacement". Even if they could not produce rain in South Australia, covering "the treeless deserts of the North" with trees would doubtless spare the colonists from the hot winds they currently experienced.[10]

But this repetition of now ancient arguments for and against trees, however heartfelt, declared Alexander Hay, was missing the point. It was the instrument of the Board, as Hogarth had pointed out, that they were to decide upon, and he personally considered this a "cumbrous and needlessly expensive" measure. In his view, cultivation of the land of any kind, including ploughing for grain crops and for forest trees, was beneficial for the climate and the beautification of the country. He believed the Government had already supplied the resources needed to complete the work of forest reserve planting and protection in the form of the public service. He considered the payment of fifty pounds a year to Forest Board members and the associated expenses of a secretary, rangers and "other paraphernalia" simply wasteful, when the staff of the Crown Lands and Public Works offices were already in place – and he was appalled that Henry Ayers thought otherwise. This "present system of

---

8 ibid.

9 ibid.

10 ibid.

multiplying Boards was just wrongheaded", he said, revealing his particular bias toward keeping the public service lean.[11]

But, countered the Chief Secretary, the field of forestry was one without established expertise in the colony and, like any new industry, would first require specialist applied knowledge before it might then pass into the normal business of Government. And so, the exchange proceeded, now with minimal dissent on the necessity of forestry, but a proliferation of conflicting views on the means and associated funding of its governance and execution.

Eventually, the floor was returned to the Bill's sponsor for summing up.[12] Ayers first dealt with Hogarth's belief in the superiority of the pure and simple man's practical knowledge and the lack of need for the expertise in forestry. He "believed knowledge was power", and wanted to see it exercised in this important colonial enterprise. He noted, with not-quite-veiled disdain, that the "great point against this Bill was the guinea a week". Where was the limit to the business of government to be set, he asked? It was clear to him that the government's work was onerous and extensive, that the district councils had been repeatedly asked and would not assume the duties of forest development or could not through lack of capacity, or simply because they did not yet exist where it was intended reserves should be. As for the suggestion of sinecures for mates, he doubted this very much, given the small amount of remuneration offered board members. Indeed, he claimed, these days "one could scarcely get a workman to put a brush of paint on one's front door without paying him a guinea".[13] But his assertions failed to prevent a few more discursive rounds about the honour or corruption of men. At length, though, Ayers was able to declare that this new Act and the Forest Trees Act of 1873 incorporated and construed as one for practical management purposes. However circular the Upper House's treatment of the bill, it did not come close to the convolutions of Lower House debate.

Two steps forward and one back: the dance of innovation

If the debate in the Upper House could be summarised as circling two issues - the practical man versus the man of science, and the appropriate governance and management structures for forestry – much wider ranging issues arose in the Lower.

---

[11] ibid.
[12] ibid.
[13] ibid.

The bill was debated continuously through the second half of 1874, but was finally allowed to lapse in the final November session.[14] The House laboured through each of the bill's very detailed clauses in turn.[15] The most vocal and extended debates concerned the governance and management of forestry. These lasted for over fourteen months. An overview of the major areas of contention offers an interesting perspective on the tactics necessary to manage innovative legislation through to successful resolution and the balance in debate between intellectual reasoning and personal value systems.

After leading with the rational economic and very local argument, Krichauff moved on to the typical centrepiece of his advocacy: an overview of what was happening on the world stage. This time, he noted that India was taking a "foremost position", while New Zealand's Government intended to introduce forestry legislation in its forthcoming session.[16] These updated examples preceded the very familiar point. If countries with abundant forests and rainfall were acting on forestry then it was beyond time for the colony to do so. He coupled the recent history of Goyder's reports on forest reserves and the impact of the 1873 Act, a member of the Upper House having recently planted thirty thousand trees on his estate. That anonymous gentleman would certainly qualify to become a member of the proposed Board, he said. This Board would be instated for the purpose of declaring boundaries for reserves and defining the regulations for planting them.[17]

Krichauff was again keen to demonstrate a disinterested advocacy, quoting and paraphrasing from a speech in the New Zealand House of Representatives. Forestry was indeed the "question of questions" for governments globally, asserted the speaker. The "industry" had the ability to extinguish the national debt single-handedly.[18] In South Australia, this was twelve million pounds. Krichauff reminded members that he was not as sanguine as either the New Zealanders or Goyder about the capacity of forestry to raise large revenues, but he did share their commitment to posterity, and believed any debt incurred for forestry to be worthwhile. Other benefits that would naturally follow – climatic, economic and aesthetic – would negate the initial expense. For him, the essential

---

[14] *SAPD*, "The Forest Board Bill. Appointment of Chairman of Committees", 12 May 1875.

[15] *SAPD*, "The Forest Board Bill", 15 July 1874; 19 August 1874; 9 September 1874; 7 October 1874.

[16] *SAPD*, 1 July 1874.

[17] *SAPD*, 15 July 1874.

[18] ibid.

significance of the activity demanded the expertise of professionals to realise systematic planting.[19] Without the appropriate leadership proposed by the bill, forestry would simply fail, he declared.

In the tedious debate of individual clauses that followed, the most contentious issue related to the powers which the bill sought to confer on a board, and the proper accountability of it to government. A hotly contested issue was the remuneration of board members. Interestingly, given the awareness of the flammability of native vegetation, the question of fire management was raised, but was also dealt with almost dismissively.

Money matters: remuneration of Board members

Existing boards, such as the Botanic Board, were used as benchmarks for remuneration. Its members received compensation of a guinea a meeting, plus sixpence to cover travelling expenses. Some members felt inclined to repeat these arrangements for the Forest Board, perhaps because of the similarity of expertise involved. Others thought the Botanic Board very different indeed. The Forest Board would have "very great responsibilities", argued Pearce, especially in their ability, on the Governor's recommendation, to "alienate certain lands". These thought Pearce, required that board members be honorary, not paid, and have "paid servants to carry out its duties".[20] Whether or not the Chairman would be one of these paid servants was highly contested. Some believed that a paid Chair would necessarily reduce the interest and commitment of honorary members. Others felt that a paid Chair would concentrate too much authority in the role, and the role should instead be interchangeable amongst members.

The debate about the standardising procedures and machinery of boards, and having the Forest Board conform to these, held the focus of members for some time. Rowland Rees, engineer, architect, and frequent member of the boards of public institutions, considered that, in time, the Board "would be second to none in South Australia".[21] Only the Central Health Board possessed a similar critical status in colonial business, and it had a paid Chairman to reflect that. Smith believed that the civic duty of many citizens would lead them to donate their time and effort for no fee or reward. Krichauff and the Chief Secretary intervened to suggest the Chairman not be paid for some years, and be elected initially by a majority

---

[19] *SAPD*, 19 August 1874.
[20] *SAPD*, 9 September 1874.
[21] *SAPD*, 7 October 1874.

of Board members. Managing the risk of the bill being voted out meant keeping expensive suggestions under careful check, as well as the time taken by their seemingly unending debate. The Chief Secretary deferred to the expertise of Goyder in the matter of remuneration for members, and would seek a report and advice about the cost and type of Board management.

Governance matters: the Board's authority

The attention of members was also drawn by a clause about the powers of the Board to fine owners of trespassing animals. Debate moved quickly to the type of animal that could do serious damage and how this ought to be accounted for in the bill, as well as the duty to fence reserved land to prevent such trespass. The obligation to fence was warranted by the severity of the mooted sanctions, which went to the extent of imprisonment for non-payment of impounding fees. Members quickly dismissed incarceration of the owners as too draconian. Smith joked that they "might as well endeavour to prevent locusts trespassing". Lindsay quipped that the difficulty would be to find the owner of the locusts. Dogs and fowl would be hard to exclude via fences. West-Erskine, in fact, wanted to advocate for the usefulness of dogs; they would keep away the more tree-damaging species of rabbits and wallabies. Horses and cattle, argued another member, must clearly be struck off the list of potential offending beasts. With debate fast descending into quibble, the Attorney-General deferred again to Goyder for advice.

Returning to the bill a month later, Krichauff announced clauses to address the issue of animal trespass. The offending animals and penalties were named. Cattle may be impounded and goats destroyed. A further amendment anticipated the obvious objection of the pastoralists. It exempted travelling stock reserves from these sanctions.[22] But rather than allay pastoralists' fears, this anticipation led to a tangential debate about the appropriate management of travelling stock reserves by the new Forest or the existing Road Board. This, in turn, led to such a heated contest that Krichauff feared the possibility of the Bill's defeat as a consequence. He asked Angas, who had gone so far as to move an amendment to the debated clause, to withdraw it – but Angas refused to submit to the limited thinking of his colleagues on the matter. He agreed to refer the matter to Goyder for advice to the House.[23]

[22] ibid.
[23] ibid.

Pearce took the opportunity of this complication to suggest mischievously that "the Honorable Member's arguments were the most conclusive that could be used to show that the reserves should be vested in the Road Boards". His view was that most of the reserves in the northern districts would be the main lines of communication, and it was equally absurd of Angas to want to hand these over to the jurisdiction of the Forest Board whatever he or the Surveyor General might think.[24]

While the value of planting the travelling stock reserves was generally affirmed for the benefit of consumers of meat in the city, the dispute over the management of the reserves and the roads of the colony had returned the debate to personalised and polarised positioning. The irresolution resulted in the Bill failing to pass into legislation in 1874. It was then left until the next parliamentary session, six months later in May 1875, for the matter to be resumed.

Bringing an end to debates

Yet again, Krichauff deployed his oratorical skill to reintroduce the large picture of forestry's purpose. Krichauff's tactic was to direct members' efforts towards the successful passage of the Bill, rather than maintain their focus on the minutiae of the clauses. The result of this particular debate, Krichauff declared, must be a small Committee to expedite the Bill and so acknowledge the majority voice in favour of the implementation of forestry.[25]

He moved towards this declaration slowly via the now standard reminder of the backwardness of the colony compared to the international examples. Close neighbour New Zealand was already profiting from an export industry in timber that had been made sustainable by forest conservation measures enacted by the government, he said. New Zealand was profiting from South Australia's tardiness to act. In 1873, the colony had imported 125,000 pounds worth of New Zealand wood, while exporting not "a single stick grown on South Australian soil".[26] The scarcity of timber and the need to expend even more on imports, not only for building but for basic requirements like firewood, should be of primary concern to members. This was even truer for inland settlers than for "maritime residents" who at least could depend upon the option of coal.

---

[24] *SAPD*, "The Forest Board Bill. Appointment of Chairman of Committees", 12 May 1875.
[25] ibid.
[26] ibid.

But remedies were immediately possible in both private and public forestry, he reminded colleagues. Farmers could plant exhausted land with trees as a crop, making their land significantly more valuable. And if public forestry management included travelling stock reserves, then pastoralists could depend upon supplies of feed from these, passing on their savings and a higher quality meat to the consumer.

David Nock, a bricklayer and general storekeeper, applauded Krichauff for his courage, after the defeat of last session to persist with the motion. With the rain that forest trees would inevitably produce, he agreed, the colony would become one of the "richest countries of the world". Quoting Isaiah 35.1, he too was in no doubt of the transformation of the colony through the impact of forests on the dry climate. It would make "the wilderness and the solitary place...glad, and the desert will rejoice and blossom as the rose". The health of colonists would benefit too as it was well known that trees would also "absorb all noxious gases".[27]

But, in spite of this inspirational oratory performance, the creation of Krichauff's Committee was thwarted. Opposing members declared that while the benefits of forests may be agreed, the means of realising them were not. Frederick Hannaford, who grew trees for a living, agreed unequivocally that conservation was essential but did not want special Boards in charge. Central management within existing government departments was his preference. On the contrary, declared Pearce, not only one, but several boards were needed, given the diversity of local contexts. With members once again polarised, Krichauff inserted himself to call for a report from Goyder. He wanted to demonstrate the large point that, in the absence of any dedicated management structure, the result of recent planting had been significant failure of the trees to thrive. Faced with this demonstrably large and ongoing loss, he hoped to inject a mood of greater compromise in the House.

A month later, Krichauff introduced the second reading, moving directly to the local situation.[28] A board was essential. The work was far beyond the capacity of the Lands Department. This was clearly evident by the poor result of last year's northern planting, where 580 pounds had been expended with few saplings surviving because of the impossibility of nurturing them. By contrast, the at Hope Valley, the proximity of which to the capital made it accessible to greater oversight, trees were growing "amazingly well, showing that all that was wanted for success was care".[29]

---

[27] Ibid.

[28] *SAPD,* "Forest Bill", 23 June 1875.

[29] ibid.

But Krichauff could not resist international comparisons for long. Continued failure to manage forests contradicted a litany of such examples. 170 miles of France's coast had been planted with trees, and the drifting sands had been arrested; the USA would alienate no further land unless the farmer agreed to cultivate ten per cent in timber trees; Switzerland's highland forests were now protected, renewed and extended following 1858 laws to stop their "reckless treatment"; pulmonary disease was rising in Hungary and Bohemia following the complete destruction of entire forests, and the ensuing winds from the Carpathians filled the air with "unceasing clouds of dust"; and Germany, Sweden and Norway had measures in place to prevent more land being cleared than was planted in annual cycles.[30] Krichauff "had no doubt that the dust [in South Australia] increased the development of pulmonary diseases" and pleaded that they not be the last country to take up forestry, when they were clearly the most needy. All that was required was this "simple measure" of "such a Board as the Governor thought proper", the payment of a small fee, the appointment of some officers, some incentives for private planting and disincentives for the destruction of trees by animals.[31]

Nock took up Krichauff's plea, reiterating the moral obligation of members to posterity. A duty to the future was owed. Policy that "did not look beyond the present time was both unwise and selfish".[32] It was just like "a man living up to or beyond his income without making any provision for the future". He returned the house to the comparison with the USA, citing a Pennsylvanian law providing that ten per cent of land should remain perpetually as woodland. If that were done by land owners for a period of ten years, an abatement of fifty per cent of the purchase price was allowed. Forfeiting land was the sanction imposed if this was not carried out. While applauding the spirit of the law, he believed that some things were the province of government and ought not to be left to private individuals. This was one.[33]

Others wanted no government interference at all, and strongly suspected the manufacturing of information about other colonial and foreign government practice. Irish-born Patrick Coglin declared that he had travelled 18,000 miles to "this beautiful and prolific country with salubrious climate-(laughter)" and taken up and worked the land. At stake, he said, was "the principle suggested that the Government should dictate to him as to what he should do or not do with certain trees on his land".

---

[30] ibid.

[31] ibid.

[32] *SAPD*, "Forest Bill", 7 July 1875.

[33] ibid.

The Commissioner of Crown Lands countered by simply pointing to Goyder, who agreed with the principles of the measure. Goyder's advocacy clearly carried a weight that appeared to need no further elaboration in the House.[34]

Lindsay wanted to challenge Coglin's view of democracy, saying, "Mr Coglin's dictum that a person should be allowed to do what he liked" with his own land just was not consistent with the "many other ways the liberty of the subject had to be interfered with for the benefit of the general community".[35] In spite of his declared interest in driving the bill into Committee, Krichauff could not resist supporting Lindsay by citing a letter from the Secretary of State of the United States. "The usual maxim of political economy which left such undertakings to private enterprise could not be applied", quoted Krichauff, because the time it took for forests to reach maturity made it "impossible for a single individual to obtain any, or more especially the highest, returns for expenditure even once in a lifetime".[36] Krichauff added pertinently that the continual turnover of governments could also lead to similar neglect of sustainable forestry management. To counter this lack of continuity, he said, "it would be better that a few men taking special interest in forest culture should be constituted a Board".[37]

Krichauff was showing impatience with the protracted need to persuade and cajole, the exercise of which he had carried the parliamentary weight of for almost five years. But governance and management issues were not to be laid to rest quickly. They went to the heart of members' beliefs about the role of government. Thomas Playford, who also grew trees for a living, introduced a different argument for a dedicated Ministry and expert public servants.[38] Hannaford thought one board useless, and wanted several. The Commissioner of Crown Lands declared that the "pressure of business on [all] Ministers of the Crown was so much at present" that the Act had to be administered by a board, assuring Playford that it would be subordinate to a Minister. Smith argued for a cheaper option. Employ Crown Rangers to do the work, he said.[39] Others argued that a the employment of a properly trained Conservator, probably by definition from England or Germany, under a specialist board and responsible to an existing Minister, was imperative.

---

[34] ibid.
[35] ibid.
[36] ibid.
[37] ibid.
[38] ibid.
[39] ibid.

Playford would have none of it. He wanted efficiency and, for him, that meant a Conservator reporting directly to the Minister. The expense was irrelevant. Moreover, if one central board was appointed would there not be an immediate outcry that local boards were needed? As if on cue, Lindsay took up the cause of decentralised management arguing that "we had adopted the Chinese policy of Boards in many directions", and as long as that system was continued, he did not think this should be made a special case. Several members argued that a central Board would be dysfunctional. How could it possibly deal with forests hundreds of miles away? Only local boards would have the local knowledge required for successful planting and conserving. Krichauff reinserted himself quickly in an attempt to deflect yet another polarised argument. He tried to assuage all parties by suggesting that a central board was only necessary to seed the innovation. He also reminded members that there were no district councils or local boards in some locations where the larger reserves were proposed. Even had there been, the benefits that would flow from forestry should flow to the population throughout the colony, and not be confined to individual areas. He would ask that Playford withdraw his amendment for a new Minister on the proviso that the Board report to an existing Minister. Playford acceded.[40]

Despite this, more animated debate followed. The extraordinary diversity of strongly held opinion demonstrated that, throughout the nineteenth century, the rights and responsibilities of land settlement were the most hotly contested issues for Australia governments and their citizens.[41] Principles of proprietorial interests were at stake, and these were exemplified in the debates over forestry. Who should have the power to allocate, use and manage public lands, were principles that sat at the heart of this Bill. This generation of leaders of the colony was returned through forestry to fundamental philosophical issues of its founding.

There was less lofty debate too, conducted in anecdotal, emotive and personal terms, rather than the profound language of complex philosophical debate. This sustained the recursive, seemingly incessant nature of discussion, sometimes at comic levels. The pastoralist George Hawker, for example, worried over the legitimacy of turning sixty acres at Mount Lofty over to the care of the Forest Board. He pointed to "one practical difficulty … which seemed to have escaped the honourable member having charge of the Bill". He had ignored its use by "large

---

40 ibid.

41 M. Williams, "More and smaller is better: Australian rural settlement 1788-1914". In *Australian Space Australian Time* (Australia: Oxford University Press, 1975): 61-103.

numbers of persons [who] went there picnicking in summer, and that he thought they would not attempt to prevent, and the trees would be destroyed, even unintentionally". Yet, in spite of the seemingly endless grounds for dissension, the bill finally passed in August 1875, thirteen months after its introduction.[42]

Krichauff as the leading parliamentary voice, Goyder as the principal civil servant, and supportive Commissioners had steered its passage together. They had navigated diligently to avoid the often imminent risk of the initiative overturning, not because of the merits of forestry itself, but because of the extent of members' vested interests, their propensity to resort to anecdotal evidence and their limited knowledge of the field. Along with patience, perseverance and repetition, they had employed a final decisive tactic: Krichauff took the extraordinary step of asking that the Standing Orders of the House be suspended to enable the Bill to pass through the remaining stages without further delay. Some asked, was this really necessary? Yes, it was, insisted Krichauff. He had reported on the Bill three times over thirteen months and considered that it was more than reasonable that it be allowed to pass. Orders were duly suspended. But even that did not stop some. John Carr, another pastoralist inflamed by the idea of managed forests, saw fit at this eleventh hour to declare he

> thought the good to be expected from this Forest Bill had been very greatly exaggerated (hear, hear). He rather regretted that there was to be another Board appointed, as it seemed the government of the colony was resolving itself into the Chinese form – having Boards of Mandarins to do the work which the Government ought to do.[43]

It was not the "stage in the history of the colony when anything could really be done", and anyway they were not going to do anything much, he mocked, except spend a lot of colonial funds. Simply setting apart small reserves around the country, and then spending a great deal fencing and protecting them, was wasteful. And, added to that, was the expense of payment to board members, when, in the end, most of the reserves were where sheep travelled. If they "were allowed to get amongst the trees for a single night it would involve replanting the whole area". It would be far better, he said, to buy up ten to twenty acres of cheap land at Meadows, if the Government insisted on making provision for forestry, where fine redgum grew "almost spontaneously". He was "exceedingly dissatisfied" with this Bill that spent a lot of money on salaries for very questionable

[42] *SAPD*, "Forest Bill", 25 August 1875.
[43] *SAPD*, "Forest Bill" 18 August 1875.

benefit to the country. Carr was supported by Coglin who believed that "those who were fond of arboriculture could grow trees themselves and pay for them" and save the colony from expensive Boards. This time it was Playford who expedited matters, scolding Carr and Coglin for not speaking up earlier and instead striving to throw the Bill out after such protracted and detailed deliberations had taken place in the house.[44] On the 25 August Krichauff was finally able to move "that the Forest Bill now pass, and the title be *An Act to make provision for the appointment of a Forest Board"*.[45] But no consensus had been achieved. It had required a procedural manoeuvre to bring to an end the endless debate of the lower house.

By 16 September, the upper house had moved through its own machinations, outlined at the beginning of the section, and the colony had the beginnings of one of the oldest plantation-based forestry services in the world.[46]

The end game?

Whatever Carr's motivation may have been in raising eleventh-hour questions of the fundamental value of forestry to the colony, and its legitimacy as government business, his interjections stand as a symbol of the seemingly unresolvable circularity of parliamentary debates on forestry from 1870 to 1875. Substantively, there had been little movement from the production of Goyder's report on proposed reserves in August of 1870 to the establishment of a small government agency to manage forestry in November 1875, with Goyder as reluctant Chair. Debate had ranged over the five years from the metaphysical value of forests to the human spirit to the detail of which specific trees suited which specific plots of land. The grand issue of the science of forestry-induced climate change and its potential transformation of South Australian settlements had been a major argument in Krichauff's armoury, strongly supported by his compatriot Schomburgk and other members of the house. The rightful claims of man, beast or trees to occupation of the land had been prominent too, mostly dividing along expected fault lines of wealth, land ownership and the strength of political voice these afforded. But such issues remained less divisive than they might have otherwise been in a colony not founded on

[44] ibid.
[45] *SAPD*, 25 August 1875.
[46] R. Robinson, "A Short History of Forestry in South Australia", https://www.pir.sa.gov.au/aghistory/dept_of_agriculture_as_an_organisation/collaborating_organisations/department_of_woods_and_forests#toc0.

philosophical principles, and thus one that framed the allocation of land within an overall picture of what a civil society ought to look like.

In 1873, Krichauff had apparently retreated from the big climatic, economic and internationally charged arguments he had presented for large-scale acquisition of public lands for public forests and settled for the modest bill of encouraging private forest tree planting. But the content and implications of Goyder's 1870 report on forest reserves continued to run in parallel with the debates about his measure. Whether born of Krichauff's genuine frustration at the parliament's failure to legislate for large scale forestry, or adopted as a less contentious but complementary means of keeping the large issues of public forest conservation and afforestation alive, his bill was a modest legislative resolution which kept the big picture of forestry before parliament.

It edged a parliamentary mood of general, albeit lethargic, support for forestry towards the highly controversial 1874 bill for a Forest Board. Krichauff gave renewed voice to his natural passions for Australia's large-scale engagement in the international moves towards innovation in forestry. But he did so with more obvious frustration and impatience than he had in 1870. Nevertheless, he presented Goyder's repeated reports on the state of forests across the length and breadth of the colony in parliament. Goyder, too, employed language of increasing urgency that had been absent in his earlier, more clinical presentation of the empirical data. His message was that systematic management was no longer optional. The resulting draft of a bill, describing in fine detail plans for the implementation of forestry, gave rise to a series of parliamentary debates that were, at their core, ones about the private and the public, or big and small government. These had assumed a substantive life of their own and shifted the attention of members away from the overarching topic of forestry. In doing so, they had added a further dimension to the wicked nature of the debate.

Inside the same dogged contests of these last debates sat the real gain of the past five years: the premise that large scale forestry of the kind mooted in Goyder's 1870 report was the proper business of government in the formation of a civil society. The Act which produced the Board in 1875 had to be expedited by invoking special parliamentary procedure. But the protraction over five years of parliamentary debate, in the days when this was extensively reported by the public press, had accomplished two important things. Firstly, it had ensured the process and principles of democratic government. Secondly, it had raised the public profile of forestry and its claims to occupation of the public land. However, the large uncertainties of climate and its relation to forests, of man's proper co-existence of the land with forests, and of the appropriate management and governance of both that had populated the five years of parliamentary debate were to be reprised and galvanised during the seven-year life of the Forest Board. Environmental innovation, it seemed, was a protracted process.

Conclusion: the case for an historical wicked problem

This historical analysis demonstrates that the problem of introducing scientific forestry in Australia constitutes a "wicked" problem in today's management literature. The broad descriptors used in the literature apply. Such phrases as "seemingly intractable", "highly complex", "social messes" and "systems of systems" can be seen to apply to the thought leadership of forestry in the parliamentary debates of 1870 to 1875.

Usually, an analysis of a wicked problem would be significantly more condensed, synthesising and extracting key themes for the reader. While this case study has rendered the narrative more concisely than its actual progress in parliament, there has been as little summarisation as practicable precisely to demonstrate empirically the abstract descriptors used in the literature of wicked problems were used in exactly the same way in this historical period. The chapter also establishes the complexities of the problem faced by Goyder as he begins the process of implementing the innovation. However, the chapter has a self-contained argument: that such problems not only existed in the past but can be described in accessible, concrete narrative that provide a beginning, middle and ending unavailable in the present. The chapter has also demonstrated the global nature of such problems, also commonly described as a much more recent phenomenon and so unable to be informed by history. Such narratives can elucidate the abstract language of the management literature of environmental "wicked" problems, and provide a concrete stimulus for the discussion of such problems in the present. What management scholars have criticised as the "presentism" of much research in their field neglects this practical contribution of history.

The empirical evidence for theory

The chapter demonstrates that, in spite of the protracted, circular and repetitious discussion of issues related to forestry, including land use, climate change and governance, resolution for action was not produced through arrival at a consensus or generally agreed upon position (even thought this may have been expected in an era that preceded present party politics). Rather, the qualities of leadership evident in producing the successful outcome were the ability to "grasp the big picture", to adapt constantly to what parliamentarians indicated they were ready for, to persevere in the face of frustration, and to hold fast to a values- and evidence-based vision. In terms of the Theory U, Krichauff's unrelenting championing of forestry in parliament indicated a leadership informed by a deep commitment to the vital importance of forestry in a sustainable colonial settlement. While the botanical and meteorological science he often depended upon to support his parliamentary oratory can be seen as flawed, and the transposition of a European view directly to a vastly different physical and political landscape equally so, he held an articulated vision of the public good firmly in his sights. This distinguished him from some of his more dilettante or ideological colleagues. His patience and perseverance seemed to be supported by this overarching vision for the colony. However, these leadership attributes were regularly tested during this initial stage, and his proclivity for arguing comparatively by using international examples was both a strength and a weakness. His vision was clearly an adopted one which, at the same time as it pushed colleagues toward adopting forestry to keep up with international trends, failed to provide a local vision of forestry or of an ideal settlement. For Krichauff, a vision for forestry was an end in itself. This was the "who" of his leadership, the deepest part of the U. For Goyder, the vision was one of forestry's place in a larger sustainable, civic settlement, and his scientific views were informed by a much more extensive empirically based appreciation of the local landscape. This case study of the thought leadership of parliamentarians begins to suggest some of the features of finding a broadly acceptable solution to an environmental wicked problem. These included a commitment to a clear vision of the public good; to the perseverance, patience and compromise needed to persuade others; and to the iterative application of the scientific method to make the best judgements possible about the local scientific validity underpinning the civic vision. The origins of this capacity are explored later. The challenges of implementing scientific forestry are explored in the next chapter.

CHAPTER 7

IMPLEMENTING FORESTRY: FEWER LEADERS, MORE INTENSE CONFLICTS

Earlier chapters have demonstrated what the environmental wicked problem of the early innovation of Australian forestry really looked like and how the leadership of it towards a satisfactory resolution began. It aimed to provide empirical evidence of for the abstractions of environmental wicked problems as described in the literature. The ill-structured, ill-defined, multi-faceted nature of such problems was shown in the messy trajectory from Goyder's initial recommendation of 1870 to the decision in 1875 to establish a Forest Board. In terms of the systems thinking that frames the book's argument, the chapter's focus was on the "what" and the "how" of the thought leadership of the environmental wicked problem, with an emerging view of the "who" of leadership. Chapter 4 reveals more sharply the "who", and distils the core elements of the "what" and "how" of leadership in the implementation phase.

The legislation to create the Forest Board meant that implementation of public forestry could now begin in earnest. But the complexities of innovation did not dissolve with this milestone achievement. This chapter looks at the evolution of the challenge of leading environmental innovation at a deeper and more concentrated stage of its development: during the life of the Forest Board from 1875 to 1882. The Board was small, comprising only five members, all with some horticultural or land management expertise, and with George Goyder as Chair. But that expertise did not mean that a diversity of views and approaches was eliminated. Rather, they became concentrated on three issues: the promise of dramatic, positive climate change through increased rainfall, the exclusive use of forest reserves for forest tree planting, and the authority of the Board to decide and act independently of government intervention.

The different type of complex problem highlighted in this chapter is that of the increased personalisation of conflict and adoption of seemingly unresolvable adversarial positions. While this element had been on display during the parliamentary proceedings, it was both more diffuse and more expected in the non-specialist culture of political debate. Within the confines of the Board, such conflict assumed threatening dimensions. The novelty of forestry, scientifically, administratively, economically and environmentally, combined with the passionate involvement of a handful of experts to render this phase of problem-solving differently challenging to its predecessor. But the conflicts were managed adaptively to produce positive outcomes in raising the public profile of forestry and helping to secure its public legitimacy.

The historical evidence used to reconstruct this phase of forestry's innovation is the extensive collection of newspaper articles produced by the local press from the outset of the Forest Board's deliberations. Goyder invited journalists to attend and report on each meeting. They did so in a style that resembles minutes, with minimal additional commentary. Other articles and letters to the editor from members of the public do provide commentary on the public reaction to and views on forestry. A key piece of primary evidence was also available: B.T. Finniss' scrapbook. This contains a selection of newspaper clippings which, taken together, reveal his interpretation of the events which led to the Board's implosion in 1882. Because of the availability of a wider range of articles Finniss' perspective is demonstrated to be a selective reading of events, compiled to favour his judgement of proceedings and to denigrate Goyder's alternative views. Published materials, including Goyder's reports as Chair, parliamentary papers, such as the Forest Board annual reports, and the Conservator, John Brown's and B.T. Finniss' published material supplement the popular record of the newspapers. In addition to these primary sources, secondary sources are also used. Researchers have written on each of the key leaders of this stage of forestry: Goyder, Brown and Finniss. Their work is used to supplement the primary sources and inform analysis.

As in the previous chapter, it is acknowledged that further primary sources could have been investigated, such as the records of government and non-government correspondence between the Board and other bodies, or the correspondence of the Surveyor General's Office. This could have been used to elaborate or confirm other primary sources. But the newspapers at this time in the colony's history reproduced accurate accounts of government activity and can be checked against each other. Journalistic and public commentary in each of the papers also provides a valuable insight into the development of public awareness of forestry, a principle goal of Goyder's during this stage of the innovation. The scope and primary purpose of the chapter is to expand the argument of the existence of complex environmental wicked problems in the past, and to demonstrate that the implementation phase of problem-solving, rather than being the realisation of the innovation, can create unforeseen difficulties that are explicable not only by differences in the "what" and the "how", but also by the "who" of leadership. The primary and secondary sources used as evidence enable the fulfilment of this scope and ensure the accuracy of the narrative.

Implementing formal forestry

Five years of parliamentary debate had produced the legislation that formalised the innovation of forestry in South Australia and Australia. But the contest of views on display in the parliamentary phase was now shifted into the new Board and given a new concentrated edge. In a small group of experts in a new field, the ideas of forestry were more deeply held and contested.

Three decisive conflicts evolved over the seven-year life of the Board, eventually leading to its implosion in early 1882. These concerned the exclusive or inclusive occupation of forest reserves, the science of the relationship of forests to rainfall, and the authority of the Board's advice to government. Once again, the neat distillation in this simple list is artificial, and disguises the messy nature of the conflicts involved with the governance and technical issues of implementing forestry.

On November 13 1875, a local newspaper, the *South Australian Register*, reported from the Government Gazette that

> the following gentlemen have been appointed members of the Forest Board under Act No, 8 of 1875:—
> Honorable Boyle Travers Finniss. J.P.
> George Woodroffe Goyder, Esq., J.P., Surveyor- General.
> Dr. Richard Schomburgk, J.P., Director of the Botanic Garden.
> Colonel William Barber, J.P.
> George McEwin, Esq., J.P.[1]

Not quite seven years later, on March 17 1882, the same paper contained the headlines that resignations had been accepted from all members. Schomburgk had expressed solidarity with the views of his fellow Board members, but could not resign his position as a government official.[2] What had happened during these seven years to cause the implosion of the

[1] *Register*, "The Government Gazette", 13 November 1875.
[2] *Register*, "Resignations Accepted", 17 March 1882.

Forest Board? The answer to this question illuminates the nature of wicked problems and how to resolve them.

The structure of the Board represented the outcome reached by parliament following its lengthy discussions on the appropriate governance for forestry. It sat outside of the formal structures of government departments, in recognition of the unprecedented work it was charged to oversee and advise on. It would report annually to parliament and seek the advice of the relevant government office where needed for the support of ongoing business.

Board members brought with them to their role a range of areas of expertise including political, botanical, and public and private sector knowledge and management, but all members shared an understanding of and experience in horticulture and/or arboriculture. Finniss and Goyder also shared expertise in surveying and an understanding of the political and public policy-making process. Finniss was appointed as the Assistant Surveyor General by the South Australian Corporation at the colony's founding. He arrived from England with a young family in 1837, and became the first premier of the newly independent colony by 1857.[3] Finniss had also led the initial unsuccessful expedition to the Northern Territory, and became its first Government Resident in the early 1860s.[4] He shared this with Goyder, who led the third and successful expedition in the late '60s.[5] His appointment to the Board was made when he was 58 years old. In the historical literature, the relationship between Finniss and Goyder has been largely overlooked in favour of a focus on the one between Goyder and Brown.[6] Finniss had a proprietorial relationship to the colony and a streak of self-righteousness that made easy the growth of divisions in the Board. The deeper reasons for this attribute will be examined in a later chapter.

The operational work of the Board was to be undertaken by the appointment by its members of a qualified Conservator of Forests, but much of this work was initially done by staff in the Lands Department. Where Goyder had suggested the engagement of his brother-in-law and locally trained nurseryman, Edwin Smith, other Board members were set

---

[3] *Advertiser*, "The Death of the Hon. B.T. Finniss", 25 December 1893.

[4] J. Booth and K. Borrow, *B.T. Finniss: First Premier of South Australia* (Adelaide: published by authors, 2001) 40.

[5] M. Goyder Kerr, *The Surveyor: The Story of the Founding of Darwin* (Adelaide: Rigby, 1971).

[6] D. Meinig, *On the Margins of the Good Earth: The South Australian Wheat Frontier 1869-1884* (London: John Murray, 1962) 71; Hirst, *Adelaide*, 53; Sheldrick, *Nature's Line*, 304.

on the appointment of an internationally trained scientific forester. John Ednie Brown was not appointed until 1878, after George McEwin had received his recommendation from noted Scottish forester, and John's father, James. At first the Board allowed leases for horticultural purposes in the reserves but within two years of John Ednie Brown's assumption of the Conservator's role in 1878, it had resolved to deny these. The government's rejection of the Board's resolution was depicted as a challenge to the wisdom in forestry science embodied in the person of Brown. It was this Scottish-born and trained forestry professional who Finniss and other members of the Board believed should hold the sole authority to arbitrate disputes of knowledge and practice in the field.

Interpretations of the Forest Board's leadership

One interpretation of the Board's demise was carefully prepared by Finniss. Amongst the archival records, a small maroon leather-bound scrapbook of selectively chosen press clippings reveals his version of the Board's journey.[7] Its focus is on the exclusive occupation of forest reserves by forests, and the corruption of a government that did not delegate absolute autonomy to the Board to decide on this exclusivity. Finniss' is a revealing account, as much for the clippings he chooses to omit as for those he includes. The scrapbook contains no personal commentary, perhaps in the hope that his simple chronological assembly of newspaper stories would be seen as an objective retelling of the truth. But however selective his choice of the evidence and its interpretation, the content of Finniss' scrapbook does reveal two of the major conflicts that emerged in the Board, the rightful occupation of reserves by humans and the autonomous public authority invested in the Board. His interpretation precedes the more inclusive narrative that emerges from the much more expansive archival record of press clippings.

According to Finniss, the split which led to the demise of the Board was one between the truths of the science of forestry and the corrupt machinations of a government that refused to submit itself to the authority of these truths.[8] One of those truths was the need for the exclusive occupation of reserve land by the forests. Human activity beyond what was needed to support and maintain forest trees must be excluded if forestry were to be given its proper place in the colony's settlement.

---

[7] B.T. Finniss papers - Forest Board, GRG 16/11, 1875 -1882, State Records of South Australia.

[8] ibid.

Finniss' clippings reveal that leasing of land in the forests began early in the Board's life; it was essential to the political economy of forestry, as the Forest Board's activities needed to be shown to be cost neutral.[9] The long-term investment model advanced in Goyder's reports had failed to persuade government to allocate sufficient funding. In order to achieve sustainable forestry, the Board believed it was necessary to demonstrate its capacity to pay for itself. Leasing was critical to this business model, as it was the single dependable source of revenue that could cover the expenses forestry would incur. The terms of leases made lessees responsible for providing the infrastructure of fencing in the reserves, thereby further reducing government expenditure. Leasing of the land also meant that the settlement of the agricultural economy could be advanced in ways that benefited individuals, society and sustainable forestry alike. The novelty of settling people and forests in the same precinct was part of the experiment of the innovation and needed to be monitored and sorted over time by trial and error, and the ongoing interpretation of the new forestry legislation.

In his scrapbook, Finniss used clippings that tell the story of one lessee. On 3 July 1876, Henry Copas applied to lease a portion of the Wirrabara Forest Reserve for nursery and garden purposes.[10] The request was initially declined, but Copas later appealed the decision. As a consequence, the Board sought legal advice in early 1878 on its power, under the Act, to resume and let lands for garden purposes. The Attorney General indicated in response that the degree of flexibility and discretion the Board possessed was almost unlimited. It was, he said, a matter of their judgement; they were responsible for interpreting the Act. In his written response, he advised them that "leasing part of the reserve for the purpose of growing vegetables, and, in fact, make market gardens of some of the land" was entirely a matter for their discretion. Should they choose to do so, the legislation gave them the power to enable "the complete shelving of the forest reserves for twenty one years at will of the Commissioner of Crown Lands".[11] He suspected that was not something they desired to do, but highlighted the extent of their capacity. In view of this advice, the Commissioner approved Copas' lease with the appropriate disclaimers, allowing for later reclamation of the land should it be needed for forestry purposes. In May 1878, the Board seemed comfortable enough with the prospect of this co-existence of forest and horticultural activity to allow for the leasing of blocks from two to five acres for garden purposes. In

[9] *GRG 16/11, Register*, 16 July 1878.
[10] *GRG 16/11*, clipping without title/date.
[11] *GRG 16/11, Register*, 17 September 1878.

September of the same year, just before Brown's arrival, eleven blocks were auctioned in Wirrabara and ten sold.[12]

Further applications for such leases, after Brown's assumption of the role of Conservator in 1879, prompted further requests from the Board for legal interpretation of the Acts.[13] One such request for advice concerned an application from an existing lessee to break up parts of his land for the cultivation of green crops. In this instance, the Attorney General's advice was unequivocal in denying the use of forest reserve land for such activity. However generous the terms of the legislation, "the breaking up of land for tillage, whether for green or other crops will incur the forfeiture of the lease" because these could not be deemed to be of the same type of activity as forest tree cultivation.[14] Growing fruit trees and forestry were seen to be sympathetic activities. This was a view shared in other parts of the new world, such as the United States.[15] The Board accepted and acted on this additional advice to exclude crop growing from the forests. In what follows, the Board seems to class only grazing as permissible, which would exclude what Copas was doing.

But by August 1880, the mood of the Board towards vegetable and fruit growing activity in the reserves had changed. Prompted by an application from a lessee who sought to transfer his garden allotments to a small market gardening enterprise, members shifted to a view that the reserves were for the exclusive cultivation of forest trees. With Goyder absent from the Chair, the Board advised the Commissioner that it intended "to remedy the earlier poor design of letting the land for garden purposes".[16] Members now believed it had been beyond their powers and the spirit of the legislation to have permitted horticultural activity in the forests in earlier years. In a report of their deliberations carried in the *Advertiser* of 22 November 1880, the Board

> maintains[s] that the Wirrabara reserve (like all others under their control) was handed over to them purely for forest conservation and culture; that they are allowed to let portions of those reserves for grazing purposes for certain fixed periods, because grazing on the land does not conflict with the real object for which it was set apart; but that to let part

---

[12] GRG 16/11, clipping without title, date.
[13] GRG 16/11.
[14] GRG 16/11.
[15] A. Wulf, *Founding Gardeners: The Revolutionary Generation, Nature, and the Shaping of the American Nation* (New York: Random House, 2011).
[16] GRG 16/11, clipping without title, date.

> of the land to gardeners is really to divert it from the uses for which it was expressly placed in trust.[17]

The Board now sought reclamation of all of the land leased for gardening purposes since 1878, including that leased to Copas. Board members started to blame Goyder's chairmanship for the poor decision-making of these early years, and increasingly found his absence from the Chair on account of competing business galling.

Eventually, members accused Goyder directly and publicly of excessive absence.[18] They deemed his dual roles as Chair and Surveyor Genera, and therefore chief adviser to the Commissioner on any contentious matters submitted for consideration by the Board, a conflict of interest. More than this, they suspected Goyder of duplicity. By late 1881, conflicts had developed to a point of irreparability. In a letter to the press, Finniss went so far as the describe Goyder as "running with the hare and holding with the hounds" at best, and as a "Machiavellian" conspirator at worst.[19] Colonel Barber resigned, and other members were discussing the option. At one of the last Forest Board meetings held on 14 February 1882, members Finniss and McEwin were reported as

> hop[ing] to bring the Commissioner of Crown Lands to reason by a threat of depriving the country of their valuable services, whereas probably this is the consummation he desires, and has been taught to desire, under Machiavellian prompting and persuasion.[20]

But by 6 March 1882, they too had submitted their resignations. These were accepted ten days later.[21]

In Finniss' scrapbook account, the Government is depicted as having treated the expert advice of Brown and the Board members disdainfully. One article compared the unquestionable expertise of Brown, his "eminent qualifications and great experience", with Goyder's lack of any credentials whatsoever. The article described Goyder as having a hubris cultivated by the government treating him as if he was "a universal genius, not simply a surveyor". The government actively encouraged Goyder, declared the writer, to think of himself as "a master of the sciences of hydraulic and marine engineering, mineralogy, and geology,

[17] GRG 16/11, *Advertiser*, 22 November 1880.
[18] GRG 16/11, *Advertiser*, 14 February 1882.
[19] GRG 16/11, clipping without title, date.
[20] GRG 16/11, *Advertiser*, 14 February 1882.
[21] GRG 16/11, clipping without title, date.

and other subjects requiring knowledge, experience, and patient toil to be thoroughly acquainted with". The writer concluded, with undisguised irony, that it was therefore only to be expected that in "so simple a matter as forest culture, Mr. Goyder should consider himself wiser than the conservator of forests himself". [22]

Finniss' scrapbook illustrates a linear trajectory of the Board's downfall. He depicts a righteously aggrieved Board unfairly and disrespectfully treated by Goyder and the government. In a unanimous resolution in November 1881, the Board (then Barber, Finniss and McEwin, with Brown and Schomburgk observing) declared their conviction that they would be "wanting in their duty to Parliament, the Public and the Government were they to acquiesce in the dictation of the Government" over the matter of the appropriate use of forest reserve land.[23] Finniss made clear that in decision-making, "we must not place any value on beliefs and opinions which do not emanate from scientific men".[24] In an undated letter to the editor of the *Advertiser*, he expressed every confidence in the ability of the Conservator to fulfil all of the duties associated with forest conservation in a post-Board future. However, he held grave doubts about his capacity to do so unprotected from the Government's "despotic control over the Crown lands". Without the intercession of some independent authority, he believed, forestry would be compromised as it had been shown to be in the resumption of lands "for their [government] own purposes". In impassioned tones, he continued that without an independent body, the entire system of forest culture would eventually be destroyed by the "parcelling out of the original forest reserves for other purposes than forest cultivation".[25] This prospect loomed large, he said, because of the legislative capacity contained in the Crown Lands Consolidation Act, no. 86, of 1877.[26] As they had witnessed, he said, this had allowed the Government to grant leases for any purpose approved by the Governor – leases which he and the other Board members now strongly believed would compromise the practice of sustainable forestry in the colony.

The following section paints a more complicated picture than Finniss draws. It draws on an extensive archive of newspaper clippings

---

[22] GRG 16/11, 14 February 1882.
[23] ibid.
[24] ibid.
[25] ibid.
[26] *Crown Laws Consolidation Act* no. 86, 1877.

kept by the Board from the time of Brown's appointment in 1879.[27] Although obviously known to Finniss, he clearly chose to omit evidence that might compromise his chosen interpretation of events.

A more complex interpretation

Brown's appointment had introduced the specialisation of professional forestry to South Australia and Australia. The appointment of a Conservator of Forests, who would be the chief adviser to government, was recommended by Goyder in his 1873 report.[28] He declared it a necessity for the successful introduction of the field.

The colony's need for sustainable forestry sat alongside and formed an integral part of its need for the sustainable settlement of an agricultural economy. Goyder had declared as much in his report on the Victorian Land Laws in 1870. The specialisation of professional forestry, the scientific credentials that added to its esteem, and the competing demands of settling a new society were played out in the life and dynamics of the Forest Board. Goyder had also made clear that the holism of social community should sit at the heart of the colony's philosophy of the survey and settlement of any of the remaining public lands. This had been the essential Wakefieldian principle of the social reformers who were the colony's founding fathers.

Locally, there was a continuing high demand for farming space in the north.[29] So pressing was this need that the wet winters of the early 1870s encouraged settlers to occupy and farm land north of Goyder's 1865 demarcation of the line of reliable rainfall.[30] The short-term perspectives of colonists and governments, coupled with their hopes and dreams for bountiful settlement, saw them ignore the wisdom that had led Goyder to pronounce the limits to growth they no longer wanted to heed. But the debates in parliament also showed evidence of a collective belief in the colony's reformist philosophy of settlement. The Wakefieldian concept of a self-sufficient farming community was one in which "the soil [would be] held and tilled by a yeomanry who would be a moral, religious, upright community spreading happiness around them".[31] A combination of the sheer will to settle and this belief in the idea of fair and equitable access to

---

[27] GRG 16/8, *Woods and Forests Newspaper Cuttings*, State Records of South Australia.

[28] GRG 16/8, *Register*, "Forest Board", 16 July 1878.

[29] Meinig, *On the Margins*.

[30] *Register*, "Pastoral" 7 September 1872.

[31] ibid.

working land contributed to the disputes between the Forest Board, Goyder, the government of the day and those settling the land. Forest reserves that were exclusive of all human occupation made no sense in the South Australian context, both socially and environmentally, was a world away from this reality.

Goyder acted to enhance the principles of democratic and ordered settlement. His capacity for exacting, disinterested and fair judgement was clearly and widely acknowledged. He understood that the messiness that accompanied inclusive decision-making was essential to the success of any innovation, including forestry. He knew that gradual and evolving ownership by the public was a necessary part of the process of settling the colony.[32] His punctuated interventions in the decision making of governments and the Board were driven by a mixture of a commitment to democratic management principles and the sheer pressure of competing business, a constant of Goyder's expansive remit. As testimony of that pressure, he had submitted his resignation in 1870, 1873 and 1878, but had been persuaded each time of the indispensability of his services to government and stayed on.[33] Goyder had reluctantly accepted the Chairmanship of the Board when it was clear parliament wanted to minimise expenses and assure any absence of corrupt land dealings. He was not inclined and did not expect to exercise a command-and-control management of the Board. He hoped for discerning shared leadership and decision-making, which had also been allowed for in the parliamentary discussions of rotating the Chair.

In the years from 1876 to 1880, the Board allowed some applications for lease for garden purposes, but refused others. Adelaide newspapers found the logic in the decision making difficult to discern.[34] By August 1880, the Board, comprising Barber, Finniss, McEwin and Brown, made clear that it would no longer be allowing leases of blocks for gardening purposes.[35] Their view was that sufficient fruit and vegetables were produced for settlers on existing blocks and that forest tree planting should no longer be interfered with by other planting. An article in the *Register* fifteen years later reveals the likely cause of the Board deciding against such leases.

Titled "Our Northern Gardens", it outlined the trajectory of Henry Copas' development of "Forest Gardens".[36] It becomes clear that

---

[32] ADB, "Goyder, George Woodroffe".

[33] ibid.

[34] *Advertiser*, 10 December 1879; *Register*, 21 January 1880.

[35] GRG 16/8, *Register*, "Forest Board" 10 August 1880.

[36] *Register*, "Our Northern Gardens", 14 May 1895.

Copas had an entrepreneurial spirit, both in identifying the environmental conditions required to produce fruit trees, and approaching influential members of parliament to persuade them of the value of his enterprise. The article described him as "the widely known and popular nurseryman of Wirrabara" who had been the first to see the suitability of the land for fruit growing.[37] Having had his first request for land refused, he planted elsewhere. Once established, he invited two local members of parliament to visit, both of whom were "agreeably surprised" by his efforts and offered to try to persuade the Board to reconsider his earlier petition.[38] This intervention led to the survey of the six blocks mentioned in Finniss' scrapbook, four of which leases were purchased by Copas. He matched his earlier horticultural success, and decided to float a company combining his other garden allotments with these "Forest Gardens". The Board, in the absence of Goyder and on Brown's recommendation, refused the transfer of the leases. Copas spoke directly to the Commissioner, and probably Goyder, and asked the Board to reconsider their decision. They did so again in the absence of Goyder and upheld their earlier view.[39] Despite Goyder's intervention to try to persuade first Barber, and then Finniss and McEwin, of the need to allow for multiple use of the forest reserve at this time in the colony's settlement, the event proved to be the catalyst of the Board's implosion. Copas went on to identify large tracts of suitable gardening land during the 1880s depression in the colony. These were surveyed as working men's blocks, a term used to describe many such blocks around the colony which Goyder had surveyed and leased to men whose depression wages could not sustain their families and who needed land on which to grow food to supplement their wages.[40] The Forest Gardens in Wirrabara had also been developed at a time when the local demand for fruit and vegetables was larger than the supply.[41] In the context of such fundamental subsistence demands on the land by settlers, the singular use of the country for forestry purposes was in Goyder's mind impossible. The single focus of a specialist Board was not sustainable. It would not best serve the public good at that time and in that context. Goyder's holistic bird's-eye-view enabled these complex judgements.

Henry Copas had shared Goyder's views on carefully managed forest tree planting as early as 1873 when the latter presented his report and recommendations to government on forest reserves. In a letter to the

---

[37] ibid.
[38] ibid.
[39] ibid.
[40] Goyder, *Report of the Surveyor General.*
[41] *Register*, "Our Northern Gardens", 14 May 1895.

editor in December 1873 Copas defended Goyder's approach to the managed planting of reserves against the claim that broadcast seeding would do just as well at significantly lower cost.[42] Having outlined his extensive experience as a nurseryman in the service of private forestry in England, Copas declared that

> I would recommend the Government to adopt the plan brought forward by Mr. Goyder, this being in my opinion, the only way of accomplishing their object. All the seeds in Europe sown broadcast would never grow into square miles of forest in a climate like this.[43]

It seemed Henry Copas shared not only Goyder's discrimination in judging the use of land for the public good, but also in adapting planting practices to a new and alien physical environment.

Members of the press also captured the dilemma faced by the Surveyor General. They empathised with his efforts to manage the competing needs contained in forest conservation and afforestation, and the continued human settlement of the land. In an article on the Forest Board's assertion that too much land had been ceded literally and philosophically to human over non-human interest, one journalist noted the legitimacy of seeking to balance these competing interests: "while the position taken by the board is unassailable, it does not follow that the land in dispute should not be let for gardening". The journalist saw the dilemma of a community needing to feed itself, while also being responsible stewards of the natural environment to be manageable:

> It should not be established as a principle that no land that will grow cabbages should be set apart for forest culture, but in this case the point is that there is no other land suitable for cabbage gardening, or the production of varied articles of vegetable growth necessary for family health and comfort. While, therefore, the board could not consistently with its special duties transfer this lease, it was quite right for the Commissioner of Crown Lands to decide that the people of the areas interested must not be denied the only means of obtaining vegetables, to say nothing of fruit.[44]

In anticipatory elegiac tones, it applauded the work of the Board assisted by a Conservator of "knowledge, experience and industry" in planting and growing hundreds of thousands of trees "not only without great cost to the

---

[42] *Register*, 16 December 1873.
[43] ibid.
[44] *Register*, "Tree Culture in South Australia", 18 May 1881.

country but in a way that will result in direct profit, independently of preventing our stores of live timber from diminishing”.[45]

As is obvious from Finniss’ account, one of the irreconcilable conflicts that led to the Board’s demise was the appropriate relationship of man and forest in the occupation of reserve land, and the authority of the Board to decide this. Finniss in particular accused Goyder of embracing a view of land use which mistakenly allowed for human activity. The Board strongly advocated a singular occupation of forest trees and the expansion of forest reserves for another reason not included in Finniss’ scrapbook: the transformation of the colony’s climate through increased rainfall.[46] This supposed capacity also acted as a further demonstration of the enormous importance of forestry to the colony and, by extension, of the importance of the Board. We have seen how the science of the dynamic relationship between forest trees and rainfall had been forcefully advocated by both Krichauff and Schomburgk in the first half of the 1870s. Now another European emigrant, the Scots Conservator, employed specifically on account of his professional credentials in forestry, was also championing the cause of forestry for this same reason. Again, Goyder chose both to allow the debate and to intervene at critical junctures when it threatened either the truth of local scientific knowledge or to damage the colony.[47]

There were two issues at the core of Finniss’ view of the Forest Board’s demise. The first concerned forestry the Board’s view, following Brown’s arrival, that reserves should exclude any other activity than growing forest trees. The second concerned the Board’s authority to make this judgement without interference from government, which Finniss believed included Goyder. Goyder’s view was, by contrast, an inclusive one. He saw forestry as part of the whole of human social and economic settlement, not as a specialist occupation to be segregated. Sustainable forestry, in the sense of both growing forest trees and as an experiment yet to be proved worthwhile, required that it be broadly owned by the community. Goyder thought and acted systemically; Finniss thought and acted exclusively and in singular terms. The first was essential to successful innovation.

---

[45] ibid.

[46] *SAPP*, “Annual Progress Report of the Forest Reserves, and Forest Conservancy Generally” no. 83, vol. 3, 1879, 1-70.

[47] *Register*, “Forest Board”, 16 July 1879.

CHAPTER 8

LEADERSHIP AND THE IMPORTANCE OF SCIENCE

The previous section highlighted how the wickedness of managing the innovation of forestry began to include not only disputes between the "what" and "how" of the problem, but also between the "who" of its leadership. Finniss' scrapbook account was clearly intended to be a personal attack on both the professional ethics and the competency of Goyder. Brown, by contrast, was presented as holding the truth of what real forestry was and how it should be implemented.

If Finniss can be seen as the Board's most vocal proponent of exclusive occupation of forest reserves by forest trees, and of the autonomous authority of the Board to decide these matters, Brown may be seen as the voice of the embodiment of the other major conflict that characterised the life of the Forest Board: the science of forests and rainfall. He was steadfastly committed to the view that rain followed the planting of forests, and held the credentials of a scientific forester to substantiate his claim. At its core, the contest of leadership between Brown and Goyder was one between the claim to absolute truth attached to the science of forestry, embodied in Brown, and the genuine uncertainty attached to the uncredentialled generalist, represented by Goyder. Theirs was a personalised case study of the esteem that was becoming increasingly attached to science through the dual impact of the scientific and industrial revolutions in the eighteenth and nineteenth centuries.

The section that follows demonstrates key elements of that conflict. At the same time, it also demonstrates the promotional benefits of the successful innovation that resulted from Brown's celebrity, his strident self-belief and the growing public profile he generated. The conflict received a great deal of coverage by the press, and the controversy added significantly to the public awareness of forestry's importance to sustainable settlement.

The search for a scientific forester

All members of the Forest Board knew that they wanted a professional forester from the outset. Goyder may have hoped that his brother-in-law and nurseryman, Edwin Smith, would be acceptable as an appointee. Smith had supported Goyder's extensive observations of the plant-able areas of the colony, and knew the landscape and its potential as well as Goyder himself. But the expertise of someone with a European scientific credential would, if appointed, undoubtedly help cement the innovation of forestry in the colony. By 1878, the Board had been struggling to find such a candidate for almost three years.[1] Advertisements had been published regularly in international newspapers without success.

If the letters to the press are any guide, the public was increasingly supportive of a professional Conservator. One letter to the editor from a self-proclaimed "Old Colonist" was concerned that mere "workmen" not be engaged, however successful they might be in the productive planting and harvesting of trees. He recognised that

> there are plenty of people in the colony who have very successfully planted orchards, vines, shrubs &c., &c, but this is quite a distinct matter from the culture of forest trees. In my own humble opinion the person most fitted for such a post would be one who has had some experience in a forest country, who has travelled through our bare districts and the bush, and who has observed with a critical eye the different soils, and the sort of timber which grow best in different localities. Soil and climate taken into consideration like will produce like the world over. I hope the Forest Board will get the best man procurable for a Conservator. The question of expense should be a minor consideration.[2]

The thoughts of this old colonist typified the growing importance of educated men of science in the stewardship of the new business of forestry.[3] Global networks of experts were influencing debate in Australia too and also contributed to public discussion through the press. Sir Joseph Dalton Hooker, the eminent botanist, explorer and close friend of Charles Darwin was, for example, in regular contact with Richard Schomburgk, Director of the Botanic Garden. Schomburgk himself was in regular communication with Ferdinand von Mueller in Victoria, whose engagement with the popular practice of acclimatization kept each man abreast of the

---

[1] *Register*, 18 November 1875.
[2] *Advertiser*, 3 December 1875.
[3] Miller, "Amateur Hour", 20.

latest intelligence on seed exchange between colonies and other countries.[4] In South Australia, Schomburgk's advice in such articles as "The Influence of Forests on Climate" lent weight to the view that forestry science was imperative for the proper establishment of forestry in the colony.[5] As in England, forestry in the colony was viewed as an amateurish endeavour by German settlers, such as Krichauff and Schomburgk, in comparison with the leading European countries of France and their homeland. Reinforcing these views from an Englishman's perspective, Dr Joseph Hooker of Kew Gardens (also a regular correspondent of von Mueller) was quoted in an article titled "The Science of Forestry" saying "forestry, which holds so distinguished a place among the branches of a liberal education on the Continent of Europe, is a subject so utterly neglected in England that we are forced to send all candidates for forest appointments in India to France and Germany for instruction both in theory and in practice".[6] With a show of patriotic pride, the journalist quoting Hooker claimed that, in South Australia, at least "we are to have a Forest Department [Board] … and a Conservator, besides". Soon, he went on, it is "to be hoped, [we will also have] an Agricultural Department and an Agricultural College, [and] we trust that the science will no longer be one of the neglected branches of education among us".[7]

But such an appointment did, in the end, take some time to realise. Given the infancy of the profession beyond France and Germany, the newness of the colony, and the distance of the colony from Europe, this is hardly surprising. Advertisements did manage to elicit some local response, but the Board decided to advise these men that it was not their intention to make an appointment until they had received word from Dr Brown, LL.D., of Stirling, Scotland, "a high authority in matters of forestry".[8] It took a further four months before word was eventually received from Brown.[9] He replied recommending his son, John, for the job. John's credentials included "practical training on a large scale both in England and North America" as well as theoretical training the quality of which was evidenced by having won the gold medal of the Highland and

---

[4] R. Home, A. Lucas, S. Maroske, D. Sinkora, J. Voight and M. Wells (eds), *Regardfully Yours: Selected Correspondence of Ferdinand von Mueller, Volume III: 1876-1896* (Bern: Peter Lang, 2006).

[5] *Register*, "The Influence of Forests on Climate", 10 August 1870.

[6] *Register*, "The Science of Forestry", 18 January 1876.

[7] ibid.

[8] *Register*, "Forest Board", 3 October 1877.

[9] *Register*, "Forest Board", 12 February 1878.

Agricultural Society of Scotland for an essay on forest conservation.[10] In demonstration of both, Dr. Brown indicated that John was, at the time of writing, professionally engaged in their joint examination of the "great forest growths" of North America. But, he added, he could be ready to sail for Australia soon. George McEwin, who had likely initiated the contact with James Brown, quickly moved the acceptance of John and sought the government's immediate ratification.[11]

With John came his father's zeal to professionalise and elevate the status of forestry on the coat tails of the growing esteem attached to scientific knowledge. Unfortunately, at the age of only twenty-six, the younger Brown brought with his youthful zeal an over-dependence on what he believed to be a universally applicable set of environmental scientific theories. Equipped with this body of knowledge, he felt little need to learn about southern hemispheric physical, social and political environments. His zeal and celebrity were contributing factors in the dissemination of the value of forestry and in the Board's demise in 1882.

Brown's scientific theory vs Goyder's scientific method

The relationship between climate change and forestry as a contributing factor in the Board's implosion was rehearsed in the first annual report presented to parliament in September 1879 and written by Brown.[12] He had been in the country for less than a year when he wrote it. His exposure to European and North American forestry, as well as a novice's enthusiasm for his profession and its evangelism led to an excessive confidence in his theories of forestry. He drafted a voluminous report, initially for the Board's endorsement. The Board meeting that discussed its content was described at length in the press.[13] The report, commented the journalist, was characterised by highly detailed lists of botanical descriptions of trees found on Brown's reconnaissance of the colony in his first nine months. Similarly fine detail was also to be found in the proposed methods for the planting and care of trees. Towards the conclusion, a much shorter section claimed the ability of forest plantations to produce rainfall.[14]

The Board had voted unanimously, in Goyder's absence, to endorse the content of Brown's draft. Goyder had subsequently seen the report and voiced his objections – both small and substantial – but did not

[10] ibid.
[11] ibid.
[12] *Jamestown Review,* 28 August 1879.
[13] *Register*, "Forest Board", 16 July 1879.
[14] ibid.

persuade any other members to retract their initial unconditional approval. As a result, at the end of the thirty-plus page report submitted to parliament, Goyder attached a one-and-a-half-page summarising appendix which declared his formal objections.[15] The difference in volume itself is symbolic of the temperaments and capacity for precision of the two men.

Goyder enumerated four objections in his document.[16] Firstly, he believed Brown had greatly exaggerated the proportion of forest to cleared land in the colony, in order presumably to encourage a view of the rampant deforestation committed by earlier settlers. Goyder suggested that Brown had deliberately misconstrued an earlier report of his to indicate that a mere one hundred square miles of timber remained in the colony, and that this remaining first-class timber represented the same kind of forest tree growth that had been so wantonly destroyed by settlers in the first forty years of the country's European occupation. Goyder corrected this environmental misinterpretation, noting that, in fact, 3100 square miles of country was timber-covered, but 3000 miles of such land produced timber of an "inferior character".[17] In short, he argued that the want of timber had not been entirely caused by the recklessness of settlers, and certainly not to the extent claimed by Brown. Brown was assuming that the same cause-and-effect relationship existed in Australia as was the case in Scotland. Goyder's extensive travel and observation of the country had confirmed for him that its natural environment was in fact incapable of producing large-scale forestry of a high quality without human intervention. Secondly, Goyder pointed out the inaccuracy of several of Brown's botanical descriptions. Goyder had a reputation for, and strong professional commitment to, precision and correctness of data. He could see the risk to the credibility of the Forest Board and forestry management of such inaccuracies, however trivial they may seem to an audience of non-specialists. Brown's reference to ironbark had been simply wrong, he said.[18] Goyder had seen the tree growing abundantly in the Northern Territory as well as in the gold-bearing regions of Victoria, but had never seen it in South Australia.

However, it was Goyder's third objection that established the significant ground of difference between his views and Brown's, both in their scientific knowledge and in the principles on which each man constructed that knowledge. Brown claimed that forest tree planting had the capacity to significantly alter the colony's climate – more specifically,

---

[15] ibid.
[16] *SAPP*, no. 83, 1879.
[17] ibid.
[18] ibid.

to increase its rainfall.[19] He brought to the earlier prevarication in parliamentary debates an unwavering certainty in the scientific truth of tree-induced climate change. In an effort to be both conciliatory and accurate, Goyder began by agreeing that forests could influence climate. He acknowledged that they were able to provide shelter, increase humidity, purify the atmosphere, produce springs, contribute fertilising qualities and subdue acidity.[20] This aside, he unequivocally dissented from the view that forests tended either to attract or to equalise the rainfall.[21] In fact, he declared, the country covered by the largest and densest scrubs and extending over thousands of square miles (which Brown could not yet have observed himself) were known to suffer from a relatively deficient rainfall. To attach the credibility of professional science to this view, Goyder invoked the evidence of "men of high scientific attainments and undoubted ability".[22] Collectively, he said, their research on the matter did not justify the conclusion arrived at in the Conservator's report. He cited in particular the findings of the British Association of Engineering, which affirmed von Humboldt's earlier findings of 1851. They concluded that felling trees on the tops and sides of mountains produced two calamitous outcomes for future generations: the scarcity of water, and the want of fuel. They confirmed, with von Humboldt, that plants exhale fluid from their leaves with various secondary effects, including the maintenance of humidity in the air.[23] They were silent on a direct causal connection between forests and rainfall.

Goyder also invoked von Humboldt to make the argument about society not science. Brown's draft report was already hinting at his view of the necessity of human exclusion from forests in order partly to maximize climatic and economic benefits. Goyder wanted to arrest this line of argument, and he wanted to do so tactically, as well as substantively – through the voice of scientific men. He knew the cultural esteem they were acquiring through the scientific revolution reinforced by the industrial revolution of the nineteenth century rendered science and truth synonymous.[24] He went straight to the European scientific authorities of the day not only to add credibility to his views on the natural environment, but also to balance the science with the social, and affirm the latter as the

---

[19] ibid.

[20] *Register*, "Forest Board", 16 July 1879.

[21] ibid

[22] ibid

[23] ibid

[24] Appleby et al., *Telling the Truth,* 15; C. Zaloom, "Mary Poovey: Interview", Public Culture 24(1) (2012): 91.

first principle on which decisions ought to be made. He agreed with the British Associations' assessment, that the question of the maintenance or removal of forests was a matter of balance and compensation. They argued, he said, the principle that wherever the progress of population settlement required that every "portion of the soil be made to yield its quota of human food" the destruction of forests was demanded.[25] But, the Association believed, there were likely to be places in every country where the "tide of population can never flow" but where a forest could flourish and where it ought to be maintained.[26] So, the British Association concluded, where human exigencies require the destruction of forests, "whether for subsistence or health", let them be destroyed; but "where neither health nor life is concerned then let a wise system of preservation be introduced and acted upon".[27] Through the use of these esteemed voices from the Old World, Goyder was declaring his own demand for sound, discerning and continuous judgement about the balance of human and forest occupation of the land, and the inescapability of values-based debate about these matters.

Finally, Goyder concluded with an objection to the conceit of Brown's suggestion that his appointment, and the work that had been done by the Board since, had turned around the misfortunes of forestry in the colony. Goyder noted simply and without rancour that the existence of the Forest Board since 1875 and the planting that had been going on in the colony for the past twenty or thirty years contradicted Brown's assertion. His remark had been "uncalled for, and should not have found a place in the report".[28]

Goyder's succinct objection to Brown's first annual report is testimony to the basic difference of principle and professional practice between him and the rest of the Board. He was guided by a principle for balanced action, a value system that was the filter through which complex and competing public good was made. Where there was a claim of both man and nature, the former must take precedence. Nor was he persuaded by the absolute truths of scientific theory, and the application of science irrespective of context. He did, however, hold in high esteem the scientific method of creating new knowledge and applying it within a balanced value system, as his citation of von Humboldt made clear.

In concluding his list of challenges to Brown's and the Board's annual report, Goyder took pains to declare that in doing so he intended no

---

[25] *Register*, "Forest Board", 16 July 1879.

[26] ibid.

[27] ibid.

[28] *SAPP*, no. 83, 1879.

personal criticism of the new Conservator. Perhaps mindful of the growing personalisation of the conflicts between himself and the Board, he distinguished between a difference in particular professional judgements, and his personal and professional respect for the new appointee. Implicitly, this declaration included the other Board members, who had supported the report. He ended by declaring his utmost confidence in the Conservator's "ability and assiduity".[29]

But members of the press seized on the appended objection to Brown's report. The *Jamestown Review* was the local paper in the young township of Jamestown, which was close to the northern forest reserves. Its criticism of Goyder was scathing. An editorial titled "Official Irrascibility" questioned his capacities in surveying let alone arboriculture.[30] He should, the paper wrote in a leader, leave the work to experts and certainly refrain from the hubris of challenge in a field in which he has no mastery at all. In fact, the public had "not an iota of reason for supposing that he has ever mastered the rudiments of forest culture" which gave him limited justification "to challenge or override on his own unsupported *ipse dixit* the matured opinions or report of an acknowledged expert in arboriculture".[31] And anyway, it went on, what gall to presume to override the Board's views given his merely nominal leadership. Surely his absence of any scientific judgement had been proven in the silliness of the line of rainfall named after him. The simple evidence of consistent rain in the years following its demarcation was positive proof "that as a meteorologist he has been a signal failure".[32] The article expressed regret at the "singularly disingenuous and most femininely illogical manner" in which Goyder had used his position "to raise to the report of the Conservator objections which he could not sustain, and to persist in demurrers which he did not even attempt to justify".[33] The matter of forestry was "a national undertaking of greater consequence than any change of policy" which must be "kept out of the reach of official incompetency".[34] The article gave voice to the more widespread view that only those with specialist credentials were the rightful deliverers of credible truth about forests.

---

[29] ibid.
[30] *Jamestown Review*, "Official Irrascibility", 3 July 1879.
[31] ibid.
[32] ibid.
[33] ibid.
[34] ibid.

Forests produce rain: a widespread belief

As the earlier parliamentary debates made clear, Brown was not alone in his view of the relationship of rain to forests. Schomburgk and Krichauff had presented papers on the subject at various public forums. As early as 1870, the *Register* ran a story on the relationship between trees and climate, reporting on an address by Schomburgk to the colony's Philosophical Society in the presence of the Governor of South Australia. Its title was "The Influence of Trees upon Climate".[35] The prospect of attracting a large immigrant population and its spread into the arid interior made the idea of trees improving the climate very appealing. The journalist reported that "the accumulated testimony in support of the view so earnestly urged by Dr Schomburgk is so unanswerable that amongst all scientific authorities controversy as to the fact of the positive influence of trees upon climate has ceased".[36] In Queensland the attention of their Acclimatization Society to forest conservancy had been reinforced by Schomburgk's paper. Their Vice-President, Mr. J.A. Bernays, expressed high praise for it citing his own view of the "sympathy between the topics of arboriculture and pluviology".[37]

In the same year, in an article titled "Forest Conservation and Forest Culture", George McEwin's view that even "the most sanguinely disposed cannot expect to prevent evils which are purely climatic…but science and everyday experience have shown that even these may be mitigated" was reported.[38] Nature itself had provided the means for counteracting climatic influences, which might otherwise have been considered to be beyond the capacity of man alone to affect. Through the drafting of legislation and the subsequent activity of the people, McEwin asserted, the "subtle sympathy between forests and atmospheric moisture" could be captured and enhanced by man to benefit the colony. He noted that the average rainfall of South Australia, some twenty-one inches, placed it "at the bottom of the list amongst Australian colonies, if not amongst the whole of the British dependencies". This was mitigated somewhat, he said, by the fact that the rain mostly fell "at seasonable intervals". Nevertheless, the "wisdom of increasing it must be universally acknowledged", especially in the northern regions of the state where there was "an immense tract of magnificent land fit for the plough but sadly, as yet, not blessed with reliable rains".

[35] *Observer*, "Forest Conservation and Forest Culture", 10 August 1870.
[36] *Register*, "The Influence of Forests on Climate", 10 August 1870.
[37] ibid.
[38] *Observer*, 10 August 1870.

Like many fellow advocates, McEwin argued his case comparatively, calling on the evidence of other countries where the rainfall had decreased considerably after the destruction of "extensive primitive forests".[39] Syria and other parts of Asia, he said, had once been "fertile, rich and populous" but had evaporated into arid wastelands as a consequence of man's environmental menace. Mauritius offered a further example of the "theory that the trees attract moisture to the earth". As an early colonist, McEwin also drew on his own observations of South Australia. Whole tracts of forest trees had disappeared, he declared, including those he named as the Black Forest, the Tam O'Shanter Belt and Pine Forest at Islington. The Mount Lofty Ranges, especially the hills' face surrounding the city, its summits and gullies had suffered the same fate, which he believed had resulted in a diminution of rainfall. At his own residence, twelve miles north east of Adelaide and approximately one thousand feet above sea level, he had witnessed a gradual decline in the annual rainfall.[40]

But McEwin did not attribute the sparseness of trees in his colony to man's destructive activity alone. He believed, like Goyder, that natural causes were the main reason for this deficiency. Principal amongst these was the large expanse of "calcareous soils" that produced a "luxuriant coating of grass and in turn prevented the growth of trees".[41] The fire-prone nature of these grasses in summer was a further deterrent to the seeding and spread of timber trees. These factors, combined with the natural absence of large forests and high mountain ranges in the interior and their ability to attract rain clouds, "exerted a powerful influence upon climate".[42] It was clearly beyond the reach of man to create high mountain ranges, he acknowledged, but we can create forests. To do so was no less than a religious obligation. "Man was commanded to replenish the earth, and subdue it", he said, "not by conquest or rapine or bloodshed, but by the spade and the plough". The result would inevitably be not only a much larger increase in the rainfall, but a source of profit to the colony.[43]

Amongst the many countries used as points of comparison in the forestry debates, Mauritius was a recurring example. Finniss had not only spent a significant portion of his childhood there, but had returned there as an army officer in adulthood. During a period of service, he had been stationed in the centre of the island in the middle of a forest seven miles

[39] ibid.
[40] ibid.
[41] ibid.
[42] ibid.
[43] ibid.

wide and treble that size in length.[44] His own experience, and that of other military men he knew who were frequent travellers through the forest, he said, was that they could not pass through it without encountering a shower of rain.

Goyder stood alone amongst the members of the Board in his views that there was no direct causal relationship between forests and rainfall. Likewise, he alone was willing to defend the claims of man's occupation of the same ground as forests. Goyder would not insist on the prevalence of his view about rain and forests. It may have been that he could see the tactical value of the publicly aired controversy that these genuine conflicts provoked as adding positively to the implementation of forestry in the colony. Whatever his motivation, the extensive coverage of the Board's activity continued, and Brown's level of dedication and enterprise contributed powerfully to the publicity forestry received, however contentious his theory of rainfall.

Brown's scientific theory in the colony's desert north

In early 1881, Brown submitted to the Forest Board and to the Commissioner a report he had drafted on research to determine the value of forest tree planting in the country north of Quorn.[45] The conclusions he drew and the recommendations he made reprised his first annual report. In it, he restated his absolute certainty that forestry would bring with it an increased rainfall while also restoring the damage done in the undoubted "indiscriminate slaughter" of the trees by settlers in earlier times.[46]

He began his report by declaring the "dearth of forest land" as the principle cause of the uncertain rainfall. With the provision of a systematic course of management, he asserted, trees could be reared with every chance of success.[47] Promotion of settlement made the provision of forests an absolute climatic and economic necessity. With the confidence of an evangelist, he claimed that "every available piece of existing forest land" should now be immediately declared a forest reserve.[48] The reserves which Goyder had proposed in his report of 1870 were to be declared and resumed in staged ways. As forestry became better understood in practice in the colony and the needs of settlers and settlement unfolded, reserves would follow. But immediate reclamation would be followed by a

---

[44] Borrow and Booth, *Finniss*, 11.
[45] *Register,* "Tree-planting in the Far North", 17 January 1881.
[46] ibid.
[47] ibid.
[48] ibid.

"systematic course of conservancy".[49] Brown was unapologetic about the urgency of his recommendations. He believed a mistake had been made in only placing timbered land under the Board after the best trees had been removed from it. This error had severely compromised its future productivity. He wanted additional unsold lands along the Willochra Creek also proclaimed as forest reserve. Finally, he recommended the laying out of the new reserve from east to west so that the "climatic influences to be derived therefrom may benefit it to the utmost extent".[50]

As with Brown's first annual report to Parliament, it was left to Goyder to append a dissenting view. With characteristic detachment, he declared succinctly the conclusions he had drawn from his own extensive experience and observation of the country, and their implications for Brown's proposal:

> The large area of the Murray Flats and other large scrubs has but extremely limited rainfall. I do not believe that trees exert any influence on the rainfall, although they keep the air moist within a certain distance of the ground, and the roots open the ground to the admission and conservation of water below the surface, which keeps up moisture to vegetation and furnishes water at lower levels in the shape of springs. Timber also prevents radiation of heat, but I do not believe that they attract or increase the rainfall. It appears to me that the ploughing, subsoiling and planting large tracts of country would involve a large expenditure, and that without irrigation or regular watering would prove a costly failure.[51]

Once again, the press weighed into the debate. One newspaper believed there was simply no contest; only one opinion, Brown's, could claim the credibility of science, and if the Forest Board considered otherwise and neglected the proper attention owed the northern districts then its reputation would be significantly diminished.[52] The reasons for initiating forestry in any part of the colony applied most powerfully in the northern districts. So, it was disappointing, "and not a little surprising", to see that Goyder was declaring it a waste of money and effort.[53] Turning Goyder's cause and effect argument on its head, the report considered that saying the Murray Flats had limited rainfall was simply a restatement of Brown's argument, and that Goyder's more moderate position that trees did prevent the radiation of heat was simply a logical step on the way to

---

[49] ibid.
[50] ibid.
[51] ibid.
[52] ibid.
[53] ibid.

their inevitable attraction of rain. The newspaper did concede that there was some evidence to suggest that settlement had been pushed beyond the limits of viable agriculture in the north but that this meant simply that caution should be taken in the experiment rather than it be dismissed out of hand.[54] The debate about Brown's report was happening as the divisions in the Forest Board were becoming more pronounced. It was both a further cause and an effect of these. So, the article concluded with a plea to the members of the Board, who felt their dignity to be under attack, to take themselves in hand and recognise that the work they were doing was too critical to the colony to allow organisational politics to impede it. But the Board did collapse in the midst of the debate, leaving this particular dispute to be played out in the columns of the press by Schomburgk and Krichauff.

An editorial published in the *Register* a month after the Board's end reported on a paper Schomburgk presented on the topic.[55] His view remained unequivocal: destruction of forests resulted in the reduction of rain, and the planting of trees *en masse* over the country was a method of increasing the annual fall of rain. He compared the meteorological records of the colony of the past ten years, showing an average increase of one inch over the previous decade's average of 20 inches.[56] Launching once more into international comparison drawn from the northern hemisphere, this time he pointed to Russian history, showing how the burning of forest cover by invading armies had had the effect of converting the country over time to desert "simply through the cutting off of the supply of rain".[57] While in France and Algiers, he noted, extensive areas had been planted with gums and other species, with the result that the quantity of rain and dew that now fell were double what they had been. Without access in the colony to the same longitudinal data these old countries possessed, argued Schomburgk, "one must take the conditions as he finds them", and plant extensive forests to boost rainfall. [58]

But Krichauff, parliamentary champion of forestry and fellow advocate then of the rainfall theory, was no longer as certain as Brown and Schomburgk. Ten days after the appearance of the *Register's* report on Schomburgk's paper, his letter to the editor appeared in the same paper.[59] He said his view had moderated over time. The causal relationship of rain

---

[54] ibid.
[55] *Register*, "Influence of Forests Climate", 15 April 1882.
[56] ibid.
[57] ibid.
[58] ibid.
[59] ibid.

and forests was uncertain, but this meant that conservative action rather than none should be proceeded with. He cited Chief Forester Schmidt from Vienna, who acknowledged the ambiguous evidence, before pleading for "further enquiry and experience". Whatever the relationship between rain and forests proved to be, he was certain that sustaining the idea and practice of forestry was an essential colonial enterprise.[60]

Leadership and missionary zeal

However questionable some of Brown's scientific assertions, his growing celebrity promoted the growth of sustainable forestry and forest culture in the public consciousness. Some of the articles in the public press began to suggest no less than a messianic role for Brown. One attributed to him the single-handed salvation of forest trees from wanton destruction. It claimed that "it is simply amazing that so unthrifty and suicidal a policy was permitted to disgrace our administration until finally the Conservator got it reversed".[61] The Conservator was "assisting nature to repair the reckless damage which reckless timber-cutters have wrought in the past". The article suggested that, had it not been for the appointment of Brown, this destructive activity would have continued indefinitely. "It is certainly amazing to think", the article reiterated, "that this wanton destruction of our forests went on without check or hindrance until the present Conservator of Forests raised an effectual protest against it".[62]

Brown could thank Goyder for the good publicity he was receiving. It was Goyder who had approached the local newspapers as Board Chair in 1877 to invite them to carry full reports on the regular meetings of the Forest Board. He wanted as much public engagement in forestry's implementation as possible to ensure its success. Following the appointment of Brown as Conservator in 1878, those meetings invariably contained extensive advice from him. It is not surprising, then, that Brown's name became almost synonymous in the mind of the public with the innovation of forestry in the colony, and his associated self-promotion did nothing to dispel this identification.

Brown's specialist status in forestry was enlisted in ways that advanced such measures as urban forestry, public education in forestry, promotion of forestry economics, and systematic measures for risk management of forests. Each of these advances in the cause of forestry is

---

[60] ibid.

[61] *Register*, "The Wirrabara Forest Reserve", 14 July 1879.

[62] ibid.

described below. That each had been a feature of Goyder's reports, the work of the Forest Board and the Lands Department well before Brown's arrival, and even foundational to planning of the city, did not discourage its identification with the celebrity of the Conservator. One of the striking features of Goyder's leadership of forestry, by contrast, was his consistent fore-fronting of the task. It is likely that he was at once disturbed by the misconceived science exhibited by Brown and pleased that the public was engaged by the Conservator's sizable ego and promotion of the value of forestry.

Finniss had sought the public disparagement of Goyder and the government in the closing months of the Board. His was a highly personalised attack on the integrity of both. As an employee of the Board, Brown was not in a position to offer the same sort of challenge to Goyder or the government, even if he was inclined to do so. Instead, his clear opposition to Goyder was made public through their polarized views on the science of the relationship of forestry and rainfall. In this instance, science, for Brown, was an absolute belief system imported from his Scottish, and extended northern hemispheric, training under his father. His science was put to service in the promotion of the widespread reclamation of land for the exclusive practice of his profession. Neither the exercise of scientific method in the dramatically different physical context of his new home nor the complex social concerns of settling a new civil society interfered with the absolutisms of his judgement. He was, however, a powerful public promoter of the innovation of forestry, even if for misguided reasons.

Brown's public education in forestry

Brown was keen to take whatever opportunity arose to promote a combination of his public profile and forestry. The Adelaide City Council's desire to introduce urban forestry in the defining parklands surrounding the city, and a practical guide to encourage farmers to plant forest trees were opportunities to do both.

At the end of 1880, Brown proposed to the Board that he further this public interest by writing an instruction booklet for the planting of trees. The primary audience would be private citizens, including farmers. With typical ambition and a touch of hubris, he suggested the title be *A Practical Treatise on Forest Culture in Australia*, even though he had direct experience of only the South Australian colony.[63] He felt that there

[63] *Register*, "Forest Board", 9 November 1880.

was considerable demand for a handbook or primer to guide tree planting not only in this but other Australian colonies, and that it fitted within the gambit of the Board's duties to respond to such demand and encourage planting in every possible way. His proposal was that he would write the book in his own time if the government would undertake to print and distribute it free of charge.

The responses of Board members varied.[64] Schomburgk pointed out, no doubt anticipating the offence it would cause his Victorian friend and colleague Ferdinand von Mueller, that the book's title should limit its audience to South Australians. Goyder was at pains to praise Brown's volunteerism and the soundness of the suggestion, while also advising that the proposed work would benefit from a deeper experience of the colony that would come from another year in the job. Mindful, too, of the political need for the Board to demonstrate fiscal conservatism in its foundational period, he asked for more information on the size and anticipated cost of the work.[65] McEwin's patronage drove a final decision; his view was that Brown's obvious ability meant that he most certainly could produce a work of substance to the benefit of the Board and South Australia.[66]

With typical enthusiasm and expedition, Brown had his manuscript ready by January 1881.[67] Brown enjoyed his celebrity status. He associated himself directly with the innovation of Australian forestry by pointing to South Australia's leadership in the field. In the introduction of his book on the subject, he claimed

> Victoria and New Zealand have each given birth to Acts of Parliament on the subject, but these have not been properly carried out. Queensland and New South Wales are now agitating in the matter; but it remained for South Australia to have the honor of being the first Australian colony to establish a system of forestry in her midst.[68]

He went on to note the growing global importance of the forestry profession in scientific circles, and himself by extension, claiming that "in the most learned savatis of the age it is now looked upon as one of the most important of the many sciences which regulate the economic and hygienic affairs of nations".[69]

---

[64] *Register*, "The Forest Board", 8 November 1880.
[65] ibid.
[66] ibid.
[67] *Register*, "Tree Culture in South Australia", 18 May 1881.
[68] J. Brown, *A Practical Treatise on Tree Culture in South Australia* (Adelaide: Government Printer, 1881): 3.
[69] ibid, 2.

The reason for the rapid production of the book becomes clear when a comparison is made between the content of the manuscript and the content of Brown's first annual report to parliament. To a large extent, Brown's practical treatise is a restatement of the report. It also incorporates a reduced, customised version of his father James' manuscript, *The Forester*.[70] A review of the first two chapters, the "Importance of Conserving Forests and Forming Plantations in South Australia" and "The Benefits which would accrue to the Soil and Climate of the Colony from a General System of Tree Planting", reveals that Brown articulates his fundamental beliefs about forestry.[71] Here, he reaffirms those elements of his first contested annual report to parliament.

Brown explored both the art and science of arboriculture. He draws on classical wisdom to highlight the wonders of Nature, and associate his work with the ages' most learned men.[72] With missionary zeal, he suggested that the supposedly primitive cultures of Asia, which had not had the benefit of a Greco-Roman foundation, were examples of the "poverty and wretchedness" that can befall peoples who neglect their moral and material welfare through contemptuous treatment of their forests. Britain, Germany and France he described as responsible for the promulgation of modern scientific methods of forestry and its adoption as an arm of industry by "landed proprietors" in these countries. South Australians now had the opportunity to align themselves with these more enlightened civilisations. Most prominent in his introduction was his reiteration of the climate science that confirmed the "well-established fact that trees exercise a very marked and valuable influence on the climatic and hygienic conditions of a country".[73]

Brown suggests the growing division in the Board between the Chair, himself and other members by minimising Goyder's role in forestry. Outlining its Australian history, he declares Friedrich Krichauff and Richard Schomburgk its true champions. He ignores Goyder's leading initiatives and, by implication, casts doubt on the credibility of his judgement and leadership, especially when compared with men of European science training. Schomburgk, Krichauff and Brown represent European wisdom, which recognised the folly of the destruction of forests

[70] J. Brown, *The Forester: Plain and Practical Directions for the Planting, Rearing, and General Management of Forest Trees* (Edinburgh: Blackwood and Sons, 1851).
[71] Brown, *Practical Treatise.*
[72] ibid.
[73] ibid, 3.

and set about their replanting many generations ago.[74] But "in these colonies", the inhabitants have been primarily concerned with the destruction of their trees (one of the points of objection in Goyder's 1879 appendix), so "unacquainted [are they] with even the rudiments of forestry". It is in the interest of their salvation and redemption from their primitivism, he implies, that "I have thought it incumbent upon myself to come forward with this concise and practical work on forest management".[75] The first part of the book continues this homily of science, modern practice, and the redemption of the primitive state of both the citizenry and the physical environment which Brown and other enlightened Europeans could offer.

Brown's pitch was aimed at the local audience of private property owners, imploring them to match their European counterparts by supplementing the work of the government and making forestry a "universal industry".[76] The wealthier citizens were judged (in spite of the failure of "Krichauff's Bill" to encourage private tree planting) most likely to respond to Brown's appeal that "every civilized country is beginning to feel that it must increase its available supply of timber". Brown believed it was they who were best placed to arrest the "indiscriminate slaughter which has marked the mismanagement of these fine and valuable lands", and it was they who were most likely to be responsive to the need for timber for domestic as well as for mining operations. This class also stood to benefit most from the provision of extensive railways and the increased timber needed to fence large properties.[77] They were also the most likely to appreciate and influence the value of urban and regional forestry, with many living in the city while managing estates in rural areas. They understood the pleasures of tree-lined streets – their shade, their purification and regulation of the air through the absorption of "deleterious gases arising from the gregarious habits of men".[78] Brown refers to his European and North American experience, including in Paris, Philadelphia, New York, Montreal and Toronto, to advise that deciduous trees are much better suited to street planting. With their annual loss and renewal of leaves comes "a degree of pleasurable variety and beauty to the scene which is not always secured from trees of an evergreen character".[79] But, above all, it was the pastoralists, the "landowners we must look (to) for that general

[74] ibid, 4.
[75] ibid.
[76] ibid, 5.
[77] ibid, 6.
[78] ibid, 40.
[79] ibid, 41.

system of planting which will ultimately ameliorate the hot winds of our summers, increase and equalize our rainfall, and conduce to the increased products of our soil". They had the funds to invest in the formation of plantations and the financial means to wait on returns from a relatively slow growing crop. The long-term material advantages of the plantations on soil productivity were also likely to be more easily envisaged by wealthy proprietors. They, too, had the sense to realise that "trees are the grand regulators of climate and improvers of the soil, under all circumstances and conditions". In his formulation, they must agree with him that

> there is no doubt but that the cause of so much waste land, so to speak, being found in the central parts of this continent, is attributable to the want of trees to give it shelter. Will anyone say that the great dry wastes at present existing in this country cannot be improved in climate and in soil by a judicious planting of proper sorts of trees on them? I believe that no one acquainted with the general laws of nature affecting such subjects will deny that these wastes may be made fertile, agriculturally, and made the field for the happy homes of thousands of settlers, by a judicious system of planting, so as to protect them from the local causes which render them at present unfit for habitation and settlement.[80]

The book then went on to provide practical and methodical advice on tree planting, as promised by the title. The publication served several promotional purposes. It was a piece of marketing collateral for forestry, encouraging the advocacy among the community of the activity and its sustainability as a government service. It was also a promotion piece for European emigration, suggesting the same civilised attention as found in Europe to both the beautification and increased productivity of the physical environment, as well as the socially innovative practices of the colony's administration. Finally, it served as a promotion for Brown himself and the scientific management of forestry, which he could be seen, through the publication, to be leading. However, while members of the Board, apart from Goyder, were unequivocal in their support of the practical, promotional and scientific purposes of the book, it received a mixed reception from the public.

[80] ibid, vi.

Public responses to Brown's education efforts

An editorial in the *Register* which reviewed the book at its launch in May 1881, urges that it is the civic duty of South Australians to pay attention to its content and act on its advice.[81] The imperative need to do so, the article said, is clear from the relative deficiencies of vegetation noted by none other than the "great naturalist Wallace" in his tour of the continent. In terms of its native flora, South Australia formed a "neutral territory" between Western Australia, with its immense variety of plants, and Eastern Australia, with its "great abundance of vegetation".[82] The editorial suggests that the only possible explanation for South Australia's natural poverty was the colony's comparative newness, which meant it had not yet planted extensively like its eastern and western counterparts. So, it was of paramount importance for South Australians to get on with the business of planting to complement its "wonderfully stimulating climate" and fertile soil. Brown's treatise was, therefore, highly welcome, and considered deserving of close attention. Its instructions were clear and succinct, able to be followed by all practical men and farmers especially. If pursued with vigour, the article went on, the Conservator promised "the amelioration of the hot winds which so frequently sweep across our open plains and scorch many a promising crop into a withered mass of refuse, and the regulation of our rainfall, which is at present too irregular and too sudden to offer anything like its full amount of benefit".[83] The commercial benefits of the timber industry to the colony were also made obvious. They were to be realised in diverse markets, not least the colony's mining community, its public works of railways and ports, and pastoralists' fencing of large tracts of land.

With the obvious strength of such advice, the article then regretted Brown's shift into much "weaker" propositions:

> Thus, for instance, in order to demonstrate the utility of tree-planting…it is not necessary for him to prove that the Central Australian Desert has acquired its arid, rainless character on account of its want of trees. The obvious reflection which occurs on reading this statement is that the writer has confused cause and effect, for certainly it appears very plain that the cause of the absence of trees in Central Australia is the sterility of the soil and the want of rain, and not vice versa.[84]

---

[81] *Register*, "Tree Culture in South Australia", 18 May 1881.
[82] ibid.
[83] ibid.
[84] ibid.

While the writer firmly disagreed with Brown's "tree theory", he applauded the eminently practical service the handbook does the farmer. A line of trees around a farm on flat land is of great value in protecting the crops from the cold winds of early spring and the hot blasts of summer. The article endorsed Brown's plea to any farmer, who has portions of his land where the soil is too poor to yield fair crops, to plant these with trees and to think of these as a commercial supplement to cropping. Wattle in particular is urged as the quantity exported is "rapidly diminishing".[85]

However, the article went on to disagree with Brown's advocacy of deciduous trees for urban planting. Their capacity for being "trained to ornamental shape" and for exerting a "purifying influence on the air by absorbing noxious gases", are contrary to trends in Europe where the Australian eucalypt is being adopted. Nevertheless, in summary, the article commended the production of the book in spite of its faults. Its distribution beyond the South Australian borders would be a credit to the colony and offer concrete testimony of the colony's "introduction of a liberal system of Australian forestry".[86] Finally, the book would at last provide an impetus for the realisation of Krichauff's desire to see tree planting by private persons and the bonus of two pounds per acre promised as reward (in fact, only two applications and one payment was ever made as a result of the Bill).[87]

Brown's forestry as moral enlightenment

An essential element of Brown's writing was the inextricable association between forestry and the demonstration of a sophisticated, moral and highly civilised people. Brown believed the social imperative of the innovation of forestry was simply a matter of fact.[88] In any country there was, he said, a direct correlation to be drawn between "the amount of attention paid by its inhabitants to their cultivation" and "the physical and intellectual standard of the people".[89]

Enlightened science, civilization and European culture formed an integrated platform of Brown's promotional strategy. American comparisons served a more ambivalent function. As another example of "newly-settled countries" America too had suffered the "indiscriminate slaughter" of forests. Brown quotes an unnamed American essayist who

[85] ibid.
[86] ibid.
[87] ibid.
[88] ibid.
[89] Brown, *Practical Treatise*, 2.

laments that "the navigation of our finest rivers is already affected by thoughtless destruction of the forests" and wonders "when will man learn that God made trees for more purposes than the lumber in them".[90]

Goyder did not publicly voice any objections to the book's contents. Perhaps his silence was recognition of Brown's success in capturing public attention for the furthering of both urban and rural forestry, not an insignificant boost to the social innovation of forestry. Perhaps it was because the Board had struggled for years to find a credentialled Conservator of Forests, and Goyder understood that the sustainability of the innovation depended to some extent on the sustainability of the management mechanism represented by the Board. The Board in turn depended upon achieving some stability in its structure and personnel. Perhaps it was because amidst the controversies about forestry, concentrated in the Forest Board, no member challenged the premise of its importance to the colony. The essential questions of the reasons for forestry, with which the parliament had grappled for years was being overtaken, however questionable some of its rationale, by its gradual dissemination in the press.

The advantage of such widespread acceptance of the significance of forestry was realised when, following the implosion of the Board in 1882, the work became the responsibility of a newly established Department of Woods and Forests. The Conservator was appointed as its Head, but the Department would sit inside the portfolio of the Department of Lands, and so ultimately remained under the direction of Goyder.[91] Forestry had finally become accepted as the legitimate business of government in managing the public lands.

Conclusion

In this chapter, it has been argued that the implementation phase of forestry's innovation demonstrated a different type of complexity to the first phase of securing its legislation. In the latter there had been a combination of broad agreement on the notion of forestry and descent into a contest of detail that could have resulted in government inertia if not for the leadership of Goyder and Krichauff. In this second phase, the contests were crystallised as rational disagreement over the science of forests

[90] ibid.

[91] R. Robinson, "A Short History of Forestry in South Australia", http://www.pir.sa.gov.au/aghistory/left_nav/dept_of_agriculture_as_an_organisation/collaborating_organisations/department_of_woods_and_forests.

causing rain, the need for specialist forestry that excluded other horticultural activity by settlers and the authority of the Board. These contests became highly emotive conflicts for the small Board galvanised by the view that forestry science, represented by the Conservator, contained the superior criteria for judgement about each of these matters.

But the chapter has also highlighted how, in this important phase of seeking broad public education in and support for forestry, the conflicts on public display in the press also generated interest in the innovation. As in the first phase, consensus was not a requirement for ensuring forestry's progress. Brown's keen interest in his own and forestry's celebrity also led to his energetic investment of time in the production of publications designed to promote the status of the field through the provision of practical advice to citizens on growing forest trees. Where Finniss's leadership was characterised by his desire to see the Board exercise absolute, expert authority in decisions related to the use of public lands for forestry, Brown had more productive aims. His comparatively youthful ego and devotion to absolutes in forestry did not, and could not afford to, contain the same destructive proclivities as Finniss. But the combination of Finniss's growing personal antipathy towards Goyder, and the Conservator's overly eager and factually inaccurate science, presented a further test of Goyder's leadership skills. He could not stem the increasing negative momentum within the Board which eventually led to its implosion, but his firm sight on a vision for forestry within the picture of a sustainable civil society, together with his use of the public press to engage the general public in the emerging field of forestry, saw the eventual positive resolution of this apparently destructive outcome. Where Finniss and Brown, with other Board members, saw their endeavour through narrow, specialist eyes, Goyder held on to a view that spanned the political, social, economic, as well as physical environment

In reconstructing the second stage of the complex problem of innovating forestry, the chapter has further demonstrated the precedents that may be found for environmental wicked problems in the past, problems typically viewed in the management literature as existing only in the present. Again, the historical context and content is obviously different to such problems in the present but their essential nature – including the contingent nature of scientific theory, the right balance between man and nature in forests, the escalation of rational difference through its personalisation to a point of irretrievability – resonate strongly with those of present problems. And the roles of leadership become more visible and more decisive of constructive or destructive outcomes.

A different type of adaptive leadership was required of Goyder during the life of the Forest Board when Finniss and Brown adopted strongly opposing positions to his claiming the superior authority of forestry science. Ironically, Goyder's opposing stance to Finniss and Brown on these issues was determined by his truer application of the scientific method, and his ability to hold on to a holistic view of the competing needs of an as yet uncertain social settlement. He did not relinquish his views, stating them firmly but in understated and disinterested ways, as in the case of his appended brief objection to Brown's first annual report to parliament. As in the first stage of innovation he understood that, however highly regarded by government and the public, his own views could not simply be imposed. He encouraged inclusive debate and experiment, recognising the need for evolution and slow emergence of sympathetic public and political opinion towards forestry. His adaptability without compromising first principles, and his disinterested but inclusive vision of forestry's place in a settling society, enabled a continuing display of committed perseverance and patience. This, together with Brown's promotional effort and the Board's well publicised conflicts, all worked towards capturing the interest of the broader public. By 1882, public forestry and public lands for its operation had become an accepted and embedded part of ongoing public and private business, and there were resources to support this. It had taken twelve years, but Goyder's ability to see his social, political, economic and physical environments as a whole, filtered through a powerful vision of an ideal settlement, eventually produced the first public department in Australia charged with public forestry.

The development of the "who" of the leadership attributes of Finniss, Brown and Goyder is examined more closely next. It will be argued that the education of Finniss, Brown and Goyder was a significant determinant of their ways of seeing and acting in the complex social and scientific "mess" of the innovation of forestry.

CHAPTER 9

LEARNING LESSONS FOR ENVIRONMENTAL LEADERSHIP: A COMPARATIVE ANALYSIS

> The prevailing system of management has destroyed our people. The destruction starts with toddlers. . . .The fundamental task of leadership is transformation of this system . . . [which is] the same system in education and business.
>
> —W. Edwards Deming, pioneer, Total Quality Management[1]

This chapter constructs a comparative investigation of the early learning environments of Finniss, Brown and Goyder based on the premise of the above quote by Edward Deming. The research of prominent systems thinker Peter Senge has used Deming's view as a starting point for research on the causal link between schooling and adult organisations.[2] He argues that the simultaneous development of formal schooling and the Industrial Revolution created schools as reflections of and feeders for industrial-age organisations, and that the present culture of learning in twenty-first century schools has not altered significantly since. His purpose in drawing on historical secondary sources as evidence is to understand and communicate the genesis of common assumptions of how schools and learning should work in order to persuade audiences of educators and organisational leaders of the need for change in the present for the future. This chapter argues that, while Deming's and Senge's generalisation about schooling has some broad-based validity, a deeper engagement with the primary sources to produce the educational histories of individual leaders during the industrialising era can also offer constructive insights for the present education of young and adult leaders alike.

---

[1] Senge, "Creating Schools," 44.

[2] ibid.

The comparative leadership of Finniss, Brown and Goyder in the second stage of forestry's innovation has been demonstrated in previous chapters. They pointed to Goyder's ability to operate adaptively within a complex and uncertain setting that became charged with internal conflict and external ambivalence, allowing the evolution of competing ideas and actions towards the acceptance of forestry as legitimate public business. It pointed to Finniss and Brown's different types of specialist approaches. Finniss had what might now be called a command-and-control view of leadership. He believed that the absolute authority for this came from the certain truths of science as espoused by the Conservator, and the structural authority he believed was invested in the Board. Brown's different leadership approach came from his professional self-belief as the only qualified scientific forester in the colony. His commitment to forestry was singular and generated an almost missionary zeal in the young Conservator. This chapter looks at the early learning environments of each man to suggest the origins of their approaches and to suggest preliminary findings of core educational principles, which may be applied in the education of present environmental leaders.

The historical sources used for this chapter are a mixture of primary and secondary. Goyder left few personal papers. Those few are personal family letters that reveal little about the private man or his personal history. But his father, David Goyder, wrote an autobiography in 1857, which offers strong evidence for the educational influences in George Goyder's youth.[3] Janis Sheldrick's literary biography of Goyder provides a comprehensive overview of Goyder's professional adult life and valuable insights into his early life.[4] The latter have been used as a starting point for further examination of the philosophies and practices of the two great influences on Goyder's parent thinking about education and citizenship, Johann Pestalozzi and Emmanuel Swedenborg. Brown's father, James, was also a prolific writer on forestry, from which much of John Brown's professional thinking can be seen as originating. Secondary sources on the Brown family supplement these publications. The most valuable source for Finniss's educational history was the biography written by Jenny Booth and Keith Borrow, a descendant of Finniss.[5] The book offers a candid and balanced assessment of the man, borne out by his own writings, as well as insights into Finniss's early life and education.

This chapter supports the argument of Senge and Scharmer that the "who" of leadership remains an under-investigated element of leadership

[3] Goyder, *My Battle.*
[4] Sheldrick, *Nature's Line.*
[5] Booth and Borrow, *Finniss.*

compared to the "what" and the "how" of leading complex environmental wicked problems. It extends the investigation of the adult leader into their early education to examine the formation of their leadership styles. But, unlike Senge and Scharmer, it uses historical methods to interrogate the particularities of individual lives to argue that the past may well be a repository of solutions as well as the location of causes of present problems. The comparative biographies reconstructed in this chapter provide historical models of the learned environmental literacies of these men that can suggest what may be critical for the present generation of students of all ages to know in order to manage themselves and others well in the natural environment.

Boyle Travers Finniss: learning to think in hierarchies and in parts

Boyle Travers Finniss' obituary in the local newspaper of 1893 begins with a perceptive piece of the man's history. Implicit is one explanation of why Finniss seemed to have lived a contradictory life. On the one hand, he exercised throughout a dutiful citizenship; he was a prominent founding father of the colony, and became its first premier after the granting of independent government in 1857. On the other, his life was frequently punctuated by conflict, often instigated by Finniss' sensitivity to wrongdoing by others.[6] His ego was fragile, and required the constant reinforcement of high profile positions. He was prone to paranoia when his point of view was opposed. Perhaps, suggested the obituary, these tendencies had begun when he was forced to follow in his father's military footsteps rather than pursue his own different interests, evident in his early schooling as described in *The Argus*:

> Mr. Finniss was born at sea off the Cape of Good Hope on August 18, 1817, and received his education at the school of the Rev. Charles Parr Burney, D.D., at Greenwich. He left there to become one of the cadets at the Royal Military College of Sandhurst. The opinion which Dr. Burney had as to his fitness for a military life may be gathered from a letter from that gentleman. Dr. Burney wrote to Mr. Finniss's father on April 6, 1822:
> —" I am, I confess, sorry that you make him a soldier, since his disposition is not very strong towards that profession, and he has real abilities which might have enabled him to distinguish himself in some of the more studious walks in life. At his examination previous to admission

[6] *The Argus*, "Death of Mr. B.T. Finniss", 26 December 1893.

> at the college he acquitted himself excellently well, though not at all better than I had fully expected. He came in at the head of the whole 16 candidates".[7]

Finniss was born into a military family, which resulted in separation from his parents from the age of four. His father served in India and then Mauritius under Colonel Boyle Travis, after whom he named his son. Stoicism and discipline seemed both part of the boy's inherited nature and subsequent nurture. So did early loss. On the ship carrying him, his mother and two sisters back to England from Madras in 1811, his younger sister died. His mother died four years later. In Kent, he was raised by a household of aunts and retired army uncles.[8]

Finniss' formal education was of paramount importance to his father. This began at a local school where corporal punishment was dispensed to children who answered incorrectly. He was quickly removed to another school with more acceptable discipline practices to complete his primary education. In 1817, at the age of ten, he was despatched to Dr Burney's School in Greenwich, not seeing his aunts again until the holiday breaks of 1821 and 1822. There, he learned the art of dealing with the physical bully.[9] With a "delicate" constitution, he became again the target of predatory boys in a culture that assumed the normality of jockeying for power in the brutal ways deemed appropriate in nineteenth-century boys' schooling.[10]

Dr Burney had inherited the school from his father. His father's school leadership had been marked by a very strict discipline, which may have been the result of his own ill-disciplined education. Burney senior had inherited the school from his father-in-law, William Rose. He had decided on a life as a schoolmaster after his preferred vocation of the priesthood was denied him by a dubious past involving expulsion from Cambridge for theft. He completed his education with a Master of Arts from King's College, Aberdeen in 1781. Included in his college years was significant "time to gamble, drink heavily, and socialize with rakish friends".[11] As a schoolmaster, Burney published on topics of classical Greek and Latin and, with some intervention from a distinguished friend, eventually redeemed himself. In 1808, he was granted a Master of Arts from Cambridge University through royal mandate, and was finally

[7] ibid.
[8] Borrow and Booth, *Finniss*, 1-9.
[9] ibid, 10.
[10] T. Hughes, *Tom Brown's School Days* (London: Macmillan, 1857).
[11] Burney Center, McGill University, https://www.mcgill.ca/burneycentre/.

ordained. His son, Charles Parr Burney, took over the school from his father in 1813, four years before Finniss' admission. The school had become one known for its education of future military and naval officers, with an English nineteenth-century mix of classical curriculum, manly brutishness and the associated vengeful religiosity of Old Testament Christianity. For Finniss, this punitive regime was not complemented as it may have been for others by the proximity of a loving family life.[12]

He seemed to learn the school's affective and academic lessons well. Finniss might have been slight but he was observant. After studying a dormitory fight where the aggressor had retreated from being struck a blow on the mouth, he mimicked the tactic when next assaulted himself.[13] This earned him a reputation of being one of the powerful boys and he was left alone as a consequence. Lessons of Latin and Greek were complemented by sports and military drills. He achieved academic as well as physical success. Though ranked in the top of his class and considered unsuited for military training by his Headmaster, his father was adamant that he attend Sandhurst. He saw to it that his son continued in his own military footsteps rather than pursuing the more intellectual life for which Burney thought him much better suited.[14]

Finniss was admitted on a cadetship to Sandhurst where he continued a quasi-military, classical training. Latin, French, maths, physics, mechanics, astronomy, theology, fortification, the new science of contour sketching, trigonometrical surveying, military drawing, riding and lancing on horseback were all part of the curriculum.[15] His fine academic results continued. Good conduct and diligence were also rewarded not only by fatherly approval but also by eventual commissioning as an officer. He was one of only five of a total of sixteen cadets to be honoured in this way. Within two years, he was promoted to lieutenant in the British army, at which time he asked to be transferred to his father's regiment in Mauritius.[16] Here, he applied his survey training in the development of roads and bridges before departing for Ireland five years later. By the end of the next five years, he had left the army, convinced that future promotional opportunities would be severely hampered by the reduction in military personnel that followed the defeat of the French.

As we have seen, Finniss' young life and early adulthood were characterised by personal loss and both physical and emotional distance

---

[12] Borrow and Booth, *Finniss*, 11.
[13] ibid.
[14] ibid, 12.
[15] ibid.
[16] ibid.

from family. Life at a public boys' boarding school and a military-centred education consolidated the emotional detachment expected of young, middle-class English males born into the nineteenth century. Despite the support of an interested headmaster who saw the scholarly potential of the young man, the dominant military culture of his family prevailed. As long as his life was framed by these military structures, it seemed to function on a stable footing, but once outside these deeply socialised modes of operating, his life became more of a pendulum swing between the assumption and exercise of his own authority, and submission to a more dominant one. It seemed almost as if Finniss conceived of his life as a power struggle that mirrored the hierarchical, military structures which had characterised his upbringing.

After leaving military life, he decided to embark on the adventure of settling a colony in the new world. He became one of the first colonists of South Australia, joining the expedition having been persuaded by the "new principles advocated by Edward Gibbon Wakefield [and] prompted greatly, not by a desire to speculate in land, but by a love of adventure and a desire to exercise my skill in the exploration and delineation of a country".[17] However, grievance and disappointment soon followed when he was beaten for the role of Assistant Surveyor General. His pride seemed unable let go of the belief that the appointment had been the result of corrupt patronage exercised by the Freemasons in his contestant Kingston's favour. Deeply aggrieved by this professional slight, he tried to withdraw from the expedition, but discovered he was contractually bound to proceed. Throughout the voyage to the colony, he continued to document the various ways in which the voyage was "beset with quarrels, poor provisioning, lack of attention to the comfort of the passengers and Kingston's domineering attitude".[18]

In his prickly sensitivity, Finniss was, at key points in his professional life, self-subverting. He would no sooner achieve a position of public esteem and potential opportunity than throw it all away. Such incidents punctuated his career, but one had a direct bearing on the abrasive relationship he had with Goyder in the implementation of the Forest Board's work. In 1864, Finniss was appointed as the first Government Resident of the Northern Territory.[19] His role was to lead an extensive survey in response to the urgent demand by pastoralists for rich grazing lands. Goyder was then Surveyor General and requested the

[17] ibid, 13.
[18] ibid, 17.
[19] ibid, 39.

expeditious survey of 250,000 acres of land for this purpose.[20] An urban centre was the first requirement. Finniss settled on Escape Cliffs. Others of his party, including a real estate agent and an *Advertiser* journalist, challenged his decision and publicly documented their contempt for Finniss' judgement. Six months later, the dispute remained unresolved, and eventually Finniss was recalled to face a three-man Commission.[21] He was found responsible for having decided pre-emptively on a poor site, which was surrounded by swamps and insufficiently above sea level to be safe. In doing so, he had spent forty thousand pounds of public money with no result.[22] Finniss again chose to see the judgement in terms of a personal vendetta.[23] Three years later, Goyder was asked by parliament to lead the third, and highly successful, expedition to survey a capital. A mere five years later, the two were thrown very closely together again as members of the Forest Board. Against this unfortunate background, Finniss was subject to Goyder as Chair of the Board.

Finniss's familial modelling and matching education for manhood had been pervasively authoritarian and emotionally cold. His biographers note that reading his letters and journals leaves the impression of an unhappy man who strongly believed that he was cheated out of the respect and financial rewards which were rightfully his. Yet, to his credit and in spite of this aggrieved state, they continue, he clung to "the sense of public duty throughout his life instilled into him during his school years".[24] In the playing out of the conflicts that characterised the Forest Board, it is difficult to imagine Finniss' trajectory being other than it was. He was schooled to see the Forest Board as having ultimate power to decide on all matters of land use where forests were concerned. He had been taught to see knowledge as absolute and external, and so was predisposed to regard science as supplying the certain truth of capacity of forests to transform climate. Man's activities should be excluded from the forest because the combined authority of science and the Board said so. Any compromise that embraced complex and balanced decision-making was, in the militaristic and hierarchical terms in which he had been schooled, likely to be demonstration of a weakness of character and judgement. Finniss' education and socialisation had taught him to be authoritarian, not a scientific authority like Brown but an enforcer of a knowledge hierarchy that had science as its head. He saw things in binary

[20] ibid, 40.
[21] ibid, 47.
[22] ibid, 49.
[23] ibid, 50.
[24] ibid, 56.

terms of the contest of right and wrong, good and bad. He held singular points of view, and pitted the might of his view against those who differed from it. He had no training to manage the uncertainties and ambiguities of the rapidly evolving colony. He had no training in learning how to learn in order to more effectively lead the emerging political, social and physical environment. If only his father had acted on Charles Parr Burney's advice that he be spared the military training to which he was not suited.

John Ednie Brown: learning to think for professional specialisation

Brown's entry in the Australian Dictionary of Biography (ADB) reads:

> John Ednie Brown (1848-1899), sylviculturist, was born in Scotland, son of James Brown LL.D., deputy-surveyor of woods and forests and an expert on European arboriculture, and his wife Jeannette, née Erskine. He was educated in Edinburgh but left school at 15 to take up his father's profession. After three years with him, learning the practical management of nurseries and reporting on forest management in England and Scotland, he was sent to the large Invercauld estate in Aberdeenshire where he learnt the profession of assistant agent and forester. He then moved to England where he laid out plantations and managed estates in Yorkshire and Sussex.[25]

Although Brown became synonymous in the eyes of Board members and many colonists with the science of forestry, his training had been practical, not academic – but it was specialised, conducted on an international scale under the tutelage of his father, the esteemed and published father of forestry in Scotland, James Brown.[26] The son's training was, as a consequence, also imbued with a highly personal and professional loyalty to an admired father's promotion of the new forestry science and its professionalism.

Brown's ways of seeing the physical environment were learned at his father's side across continents.[27] From his practice and reports during his early years as the first Conservator in Australia, what he appears to have learned best was how to insert his new Australian environment into

---

[25] R. Refshauge, "Brown, John Ednie (1848-1899)". Australian Dictionary of Biography, National Centre of Biography, Australian National University (1969).
[26] J. Oosthoek, *Conquering the Highlands: a history of the afforestation of the Scottish uplands* (Canberra: Australian National University Press, 2013).
[27] Refshauge, "Brown, John".

the theories about forestry transported from his northern hemispheric training. He had not so much learnt the underlying processes of a scientific method as the content of its conclusions in a quite different environmental context. The foundations of Brown's scientific worldview begin with an overview of some of his father's professional endeavours.

James Brown wrote the first edition of his book *The Forester* in 1847 while forester on a private estate, Arniston, in Midlothian, Scotland.[28] The book's combination of civilising and practical purpose was declared in the introduction to a second edition published in 1851. It aimed to make "noblemen and gentlemen . . . truly practical foresters, as they now are practical farmers".[29] It ran to five chapters over 215 pages. The title indicated that it was "a practical treatise", a rehearsal of his son's handbook three decades later. "Diffusing . . . a sound and extended knowledge of arboricultural operations" was James' and later John's, professional ambition. This edition contained seven chapters and an appendix divided into 100 numbered sections, and ran to 526 pages.[30] The third edition of 1861 was even more voluminous and more ambitious. Brown had, by then, left Arniston for the position of forester to the Earl of Seafield and Surveyor of Woods in General. His book had grown to thirteen chapters divided into 136 numbered sections. A new first chapter claimed for forestry the status of a profession. It began with a history of forestry, outlined its importance to the economic, naval and general welfare of Britain, and described the attributes of a competent forester.[31] Compared to the science of gardening, Brown wrote, forestry had languished in neglect, in considerable part "from the want of a properly qualified class of men having been put in charge of woodland property", and in part from the absence of a periodical literature devoted to the science of forestry.[32] The profession was still in its infancy.[33] Foresters must have their own professional training regime, just as surgeons have theirs in "the anatomy and constitution of man". Without science, it is "impossible" for anyone to be a forester. He did not know of more than a dozen "really qualified men". The rest were simply "workmen" with skills confined to planting trees and then chopping them down. Improved education was key to increasing the number of professionals.[34] The

[28] *Register*, "Forest Board", 12 February 1878.
[29] Brown, *Forester*, viii.
[30] ibid, vi.
[31] ibid.
[32] Brown, *Forester*, 6-10.
[33] ibid, 27.
[34] ibid, 34.

forester should learn arithmetic, English grammar, "all the common branches of mathematics", French, botany, vegetable physiology, geology, chemistry and entomology.[35] With an education of this sort, a forester may expect an income equal to that of "most clergymen and country doctors of medicine".[36]

In 1854, three years after John's birth and between the second and third editions of Brown's book, Scots foresters organized their own professional association, the Scottish Arboricultural Society. Brown was its first President from 1855 to 1857.[37] In a later Presidential Address in 1873, Hugh Cleghorn, physician, botanist, and pioneer of forestry in India, remarked on the shift in papers presented to the Society since its inception eighteen years earlier. At first, they had been devoted to "purely practical subjects"; now they concerned climate, acclimatization of species, improved instruments of measurement, Indian forestry, "and collateral branches of science".[38] His remarks were reported in the Society's own journal. James Brown, recently honoured with a doctorate, was still a member of the Society, as were his two sons.[39] All male members of the family were now enlisted in the profession and the job of elevating its status.

John's childhood education, like his brother William's, was received at the local parochial schools in Edinburgh.[40] The system of parish schools in Scotland was an object of patriotic pride, providing as it did a broadly based education for the urban working poor.[41] This system was pointed to favourably by those who advocated state provision of schooling in England in the early nineteenth century. Claims of "widespread literacy, economic progress, and a contented 'peasantry'" were made for the system in which both Brown brothers were educated.[42] Each stayed on beyond the legal departure age of thirteen, but the Browns did not belong to the social class that sent sons on to the university.

Instead, each followed their father into an apprenticeship in forestry and in the mission of professionalising it through education.

---

[35] ibid, 37-39.

[36] ibid, 40-1.

[37] ibid, 42.

[38] *Transactions of the Scottish Arboricultural Society* 7, 1873-8, 115.

[39] ibid, 116.

[40] ibid, Appendix A.

[41] "William Brown, Canadian History", https://electricscotland.com/history/canada/brown_william.htm.

[42] R. Anderson, "Education and the State in Nineteenth Century Scotland", Economic History Review, 36(4) (1983): 518–534.

William's destiny was to immigrate to Canada and become a professor. In 1871, he moved his family of six children to farm and survey land in Ontario. By 1875, he had been appointed to a professorship at the Ontario Agricultural College and Experimental Farm. He published academic papers including "Claims of Arboriculture as a Science", read before the British Association for the Advancement of Science, and won gold medals of the Highland and Agriculture Society of Scotland, as well as of the Scottish Arboricultural Society. He went on to lead the Guelph Agriculture College "whose reputation for excellence is now so well established", one observer remarked, "that its membership each year includes students from the most distant parts of Canada, from Great Britain, and from the United States".[43]

John was a boy of 13 when his brother began making such strides in his career, and the realisation of his father's hopes for their shared profession. It is little wonder that he left school earlier than William to join in the family mission. But his father continued to reinforce the message that education and scholarship were necessary for the proper accomplishment of the profession of forestry. John's contribution of his own papers after visiting and researching in the United States and Canada in 1871-72 was a promising sign of emulating his brother's scholarship and contributing to his father's educational ambition. His "Report upon Trees found in California" and "Forests of the Eastern States of America" also earned him the gold medal of the Highland and Agricultural Society of Scotland. Another essay, "The Trees of America", won him the prize of the Scottish Arboricultural Society.[44]

These three Scottish-born foresters represented a new kind of professional. With their knowledge of science, their professional association and their learned journal, they were confident specialists in a period which esteemed both science and industry. James had been an entrepreneur of forestry knowledge on behalf of its professional elevation, and had readied himself and his sons to take over the management of forests from the artisan and the amateur. John's training in forestry was therefore intimately entwined with the loyalty of a son to an esteemed father. James' writings on forestry and his close involvement in establishing the academic and professional standing of the field coincided with John's birth and development to manhood. Practically, professionally and emotionally, John and William had been apprenticed to their father. And because of the strength of James' commitment and advocacy of the

---

[43] ibid, 1.

[44] Refshauge,"Brown, John".

new science, father's and sons' sense of self was intricately linked to the celebrity of pioneering this new field. As his father's protégé, he had learnt his craft and his science working in countries more socially and politically mature than the fledging colony to which he emigrated. With all the undoubted knowledge he brought with him, he had also packed an unquestioning and universal belief in what he knew and from whom. He was a young man on an evangelical forestry mission!

In a colony of the New World this uncritical, portable belief in the theory and practice of his profession and its knowledge base amounted to youthful scientific over-confidence. It aligned with and reinforced the uncritical belief of Finniss in the absolute expertise and authority of the Forest Board. For different reasons, they shared a condescension towards the amateur in government and the public. Each had grown up with a wealth of international exposure but, as the earlier narrative shows, had not learned how to see and learn outside of their different unquestioned belief and value systems. They were committed to their different versions of absolute certainties and the power that gave them in decision-making.

Brown's significant contribution to the innovation of forestry was undoubtedly the pride he had learnt in the profession and a capacity equal to his father's in generating publicity, celebrity and esteem for forestry. Both were assets in the fledgling stages of innovation in South Australia in the 1870s and 1880s. As we have seen, even the singular views he held about forestry and the public controversies these encouraged worked in the service of raising the profile and importance of the field amongst colonists. But the adoption of his father's missionary zeal blinded him to the realities of his new physical, social, political and economic environment and compromised his judgement.

Summary of Finniss' and Brown's learning environments

This section has suggested reasons for the commitment of Finniss and Brown to certainties in forestry. Each man was a dedicated champion of the innovation of forestry in the colony who contributed significantly to its successful innovation. But each also made that eventual success problematic. They each saw part of the whole of forestry, and also saw forestry as a self-contained whole, not part of the larger picture of the settlement of a civil society. The section has argued that each man's partial view was influenced by an early education that promoted authoritative and absolute certainties. In Finniss' case, this was the decision-making authority of the Forest Board. In Brown's case, it was the decision-making authority of scientific theory. Neither man was invested with the capacity to learn how to learn or learn how to think about the changing nature of their external reality.

That Goyder recognized the opportunity Brown's predispositions presented, even as he knew the inaccuracy of his scientific judgements, was one indicator of his different education and worldview. He had learnt that environmental knowledge was not an absolute and unchanging thing but a continuous, iterative process of learning for understanding that depended greatly on context. And he had learnt that political, economic and social context needed to be integrated with environmental knowledge to achieve a successful outcome. The following section suggests that, like Brown and Finniss, the origin of his more holistic leadership was evident in his early learning environment.

George W. Goyder: learning to think in wholes

The argument of this section depends heavily on the inferences that can be drawn from an examination of the primary influences of Goyder's education. These were his parents and, through them the Swiss educator Johann Pestalozzi, and the Swedish government science policy adviser and mystic, Emmanuel Swedenborg. As previously indicated, Goyder left no personal papers, and tended to declare himself professionally only when essential to do so. We rely on a reasonable inference drawn from his father David's autobiography to reconstruct his childhood and education.

David Goyder was an ordained but unpaid Swedenborgian, or New Church minister and, with wife Sarah, a paid Pestalozzian teacher (the first Swedenborgian Church was founded in 1844 in South Australia; it was a likely attraction for George Goyder who remained a practising member throughout his life).[45] In his autobiography, David distilled the essential purpose of the innovative method of instruction that Pestalozzi had designed. He described how

> it was the aim of Pestalozzi to combine the powers of the understanding with the will-of thought with affection-and to bring the both into actual existence on life. Hence, his system is one of faith and love, or, in other words, he united the cultivation of heart and understanding with the labor of the hand. His motto in education was -Heart, Head, and Hand. Science he called in as an auxilliary, He contended that what was done for the head alone destroyed the heart; but what was done for the head, through the instrumentality of the heart, preserved both. Thus he laid the

[45] Hilliard, "Unorthodox Christianity".

> foundation of his pupils, and then he had the power to direct them without difficulty.

The quote describes the interior system that framed the Pestalozzian method. David was one of its powerful advocates in the United Kingdom in the first half of the nineteenth century, and his children were of course schooled in it. The quote anticipates the twenty-first century focus of some systems thinkers on the personal attributes of leadership. Education would be an integrated learning of the rational, emotional and experiential knowledge of the particular individual.[46] This chapter argues that George Goyder was a systems thinker in the nineteenth century long before the term was coined in the management and leadership literature. Through his education, he developed a capacity to conceive of complex phenomena as parts of a whole, joining up external parts of a system and linking these through his internal framework. He was not a linear cause and effect thinker, nor an exclusive or specialist thinker like Finniss and Brown. He thought, valued and observed his environment in ways described in the quote by his father and Pestalozzi as head, heart and hands.

Janis Sheldrick's comprehensive research explored elements of Goyder's professional and personal lives, introducing readers for the first time to the connections that could be made between the two. Her research uncovered father David's 1857 autobiography and through this the powerful influence of Johann Pestalozzi and Emmanuel Swedenborg in both David, and his wife, Sarah's lives.[47] Sheldrick noted the dearth of personal archival sources left by Goyder. For reasons of pressure of work, of the propriety that went with being a public servant, of family, or of personal proclivities, Goyder left very few private papers and expressed a personal view about professional matters rarely. Although an iconic figure in South Australian environmental history, such a perception rests almost exclusively on the definition of a line of reliable rainfall in the colony a mere four years after his promotion to Surveyor General in 1861. It is a popular view that reinforces the idea of the exceptionalism of the entrepreneur or innovator, whether social or corporate.

A recent example of this popular notion of Goyder's exceptionalism was the launch of the Goyder Institute in 2010. Federal and State Ministers identified him as an exemplar of someone whose science had informed environmental policy. Such a portrait has undoubted legitimacy, but it implies a view that has little instructional value for a twenty-first century

[46] Meadows, "Dancing".
[47] Sheldrick, *Nature's Line.*

audience, except as a symbol of connecting science and policy.[48] Goyder's significance in the present and the future of environmental wicked problem - solving is more substantive than this. It has to do with who he was as a leader of not only science and policy but of the settlement of a civil society. His leadership role in forestry is an opportunity to take Sheldrick's findings even deeper to explain more about the formation of this important systems thinker of the past and show how his life can contribute to aspiring systems thinkers of the future. A close examination of Goyder's education provides not only an insight into the relationship between his youthful and adult ways of seeing his world. It also offers some transferable lessons in environmental literacy for the twenty-first century leader of the management and resolution wicked problems.

Goyder was a man whose focus was fixed on the present for a sustainable future. But his ability to learn continuously was developed in the eighteenth century. The Pestalozzian method was developed in direct response to the deep political and social turmoil of eighteenth-century Europe. So fundamentally challenging were these revolutionary times that they led to a profound interrogation of the first principles of knowledge and learning for life by the Swiss educator, Johann Pestalozzi. Pestalozzi's work in turn owed something of its philosophy to the other major influence in Goyder's formation, Emmanuel Swedenborg. Trained as a scientist in the early eighteenth century, Swedenborg served throughout his life as an adviser to the Swedish parliament. In this role he combined that scientific training with his subsequent involvement in spiritual mysticism to continue to provide very pragmatic policy advice. This chapter looks at the influence of Goyder's parents and, through them Pestalozzi and Swedenborg, to suggest how he operated as a leader using head, heart and hand. More importantly it argues that Goyder's education offers historical evidence for and a replicable case study of the fundamental educational principles of systemic environmental literacy for the twenty-first century.

Before doing so, three extracts provide an expanded view of the adult Goyder. Each offers a short insight into his "head, heart and hand" leadership.

---

48 "Research Hub Secures SA Water's Future", Joint Media Release, Senator Kim Carr, Commonwealth Minister for Innovation, Premier Mike Rann, Hon Paul Caica Minister for Water, May 2010.

A summary of Goyder's leadership principles

This obituary appeared in the local press following Goyder's death at the age of 72 in 1898:

> Fortunate are the Governments who attract to their service men of the high caliber of the late Mr. Goyder....For many long years he bore one of the heaviest responsibilities in the State – the administration of a great department, intimately connected with the most important of the material interests of the people....For himself, he won honor, respect, and almost unlimited trust....His name became a household word in South Australia. It was impossible that he should escape criticism, often adverse; for he decided innumerable questions that touched private interests, and adopted methods of administration, and indicated lines of policy, on which controversy naturally arose. But there was universal confidence in his absolute independence, in his resolute integrity, and in the general soundness of his judgement....Deliberate and systematic, with a strong sense of order, he showed how well he understood the true economy of force. His industry was supplemented by organizing skill and administrative capacity, professional ability, and decision of character....Year after year...he was the official head of the land department, and the guide and counselor of Ministers, many of whom took up their portfolios raw and inexperienced, and left their offices informed and disciplined by him.[49]

The obituary demonstrates the constancy and continuity of Goyder's oversight of the settling colony compared with the continual changeability of parliamentarians to whom he reported. It highlights the integrity as well as competence of the man in the face not only of complex matters of judgement but the personal attack that often accompanied it. It also makes clear that, in the context of the complexities and uncertainties that were constants of the nineteenth-century settlement, Goyder oversaw the innovation of forestry. He could not ignore the views or the claims of settlers in decisions about forest reserves, as Finniss had wished. Nor could he afford on behalf of these settlers to invest any credibility in what he knew to be the faulty science connecting rain with forests, and so put people's livelihoods in jeopardy. Working forest reserve land for horticultural reasons, providing food for colonists and small economic enterprises for the land's lessees without causing damage to the development of the surrounding forests, was a necessary co-existence at this time in the colony's history. Endorsing the theory that forests changed the climate could have led again, as it had in the mid-1870s, to the false

---

[49] *Advertiser*, "George Woodroffe Goyder", 4 November 1898.

hopes of farmers acquiring and successfully working land beyond the line of reliable rainfall. The obituary encapsulates Goyder's capability and proclivity to operate across the boundaries of scientific, social, economic and environmental matters.

Two indicators of a practical heart

Goyder's adult synthesis of head, heart and hand can also be seen in heartfelt snippets of his otherwise clinical thirty-six-page summary of the allocation of lands from 1836 to 1890. This succinct and exacting retrospective of the disposal of public lands contains glimpses what might be called his principles of citizenship. He refers to the creation of workingmen's blocks in the 1880s, a time of economic depression in the colony. He also describes briefly the establishment of Aboriginal Reserves.

The survey of workingmen's blocks and townships occurred in areas considered to be of poorer quality soil and so of limited desirability to horticulturalists. One such survey occurred amidst the native forests of Mylor in the Adelaide Hills, close to where Goyder himself owned and ran an experimental farm. Workingmen's blocks enabled families to lease an area of twenty acres from the government on which to grow food and keep animals, supplementing the wages of salaried men whose earnings had become inadequate to support a family.[50] In the following extract from the report Goyder's language contains genuine empathy and respect for the dignity of these lessees:

> There are several persons holding lands of this class personally known to me as neighbors, who are hard-working, trustworthy, and very intelligent men; these find ready employment when not engaged upon their blocks. Others have a sufficient area of cultivable land as to require all their time; but it would be better if, even in the hills, a few country residences were scattered about where the working men might at certain times obtain remunerative employment. It is an uphill game for many of the poor fellows.[51]

There is no paternalism or condescension here, but there is considerable fellow feeling with the plight of these families and desire to preserve the esteem that accompanies self-sufficiency. He may have been the owner-occupier of one of these "country residences", but in his view that situated

[50] J. Hirst, *G.W. Cotton and Workingmen's Blocks,* B.A. (Hons) diss., University of Adelaide, (1963).

[51] Goyder, *Report of the Surveyor*, 3.

him in the social system in a particularly responsible, not superior way. His role was to enable and promote the "remunerative employment" of his new neighbours, including by appeal to his existing neighbours to support the dignity of their self-sufficiency.

Another example of Goyder's empathic approach is seen in his description of land surveyed as "Aboriginal Reserves". An immediate contemporary response may be to see reserves of any kind as anathema to an absolute obligation for respectful inclusivity and reconciliation within the operation of the dominant culture. But Goyder offers a perspective that challenges this view at this time in history. In fact, it resonates with the spirit of cultural respect on which contemporary notions of reconciliation are built. He had surveyed over one hundred thousand acres of land in various parts of the colony for these reserves. It becomes clear that their primary purpose was not racial segregation, but the enabling of the conduct of a different sustainable way of life for Aborigines in concert with traditional Indigenous culture. He outlines an insightful understanding of this culture. His tone is regretful that the measure is required because of the fundamentally western and antithetical view of the land that the Europeans had brought with them to settlement:

> These lands should be made the resort of all the aboriginals whose love of the country round their birthplace does not form an actual bar. These lands possess all that can be required by aboriginal life. There is timber, grass, herbs, shelter, game, waterfowl, fish, and water; in fact, a black turned out on a blanket and tomahawk and a few lines and hooks can get a living at once, and no further portion of these lands abutting on these lakes, or on the Coorong should be alienated, but held inviolate so long as they are needed, and there are aboriginals to resort to them; nor should the timber be destroyed as the grubs for bait are obtained from them. There are salt and freshwater fish, from cod to coongulty, crayfish, cockles, mussels, wallabies, kangaroos, and smaller marsupials; turkeys, geese, ducks &c., too numerous to mention, as well as swans and pelicans. I cannot place too great an emphasis upon the value of these reserves, and trust that they will be properly cared for, and as the landholders who reside thereon show them the greatest kindness and attention, as well as find them work when they wish to be employed, careful oversight on the part of the trustees appointed or acting in their behalf being all that is required to ensure their comfort and happiness, as far as such can be attained.[52]

[52] ibid, 44.

Goyder's ability to inhabit the ways of seeing Aboriginal cultures had also been evident in his successful survey of Darwin in the Northern Territory in 1869.[53] It was in the same spirit of preserving such distinctive cultures that Goyder describes the rationale for Reserves. This powerful capacity for empathy had its origins in how he had learned to combine intellectual insight, fellow feeling and the use of his senses and body.

It is not surprising that, in a pioneering, uncertain environment, where water scarcity for most of the first generation of European settlers presented an alarming novelty, Goyder became almost exclusively associated with the science of defining the line of reliable rainfall.[54] But Goyder is unlikely to have called himself a scientist. Before immigrating to Australia in 1848, at the age of 22, he had trained as an apprentice in the production of precision survey instruments and as an engineer in urban environments.[55] He might well have answered though to the description given him by Michael Williams as a "practical geographer" and acknowledged his expertise in natural resource management, acquired through keen, almost relentless, observation.[56] Sitting alongside the reading of his environment were the diverse responsibilities he exercised, and the social and technological innovations he conceived over a long career as Surveyor General. Often, as we have seen, in spite of exposure to public ridicule, he persisted with a quiet confidence in what environmental historians Libby Robin and Mike Smith call his "ways of seeing".[57] He apprehended the complexity of his surroundings, to realise innovations in areas including water and forest resource management, strategic land survey and use, railway services, mineral exploration, and experimental farming.

Sheldrick has outlined the breadth of Goyder's portfolio, highlighting the extent of his environmental insight and the social and political context in which this was played out.[58] She nominates seven areas of Goyder's innovative endeavour before concentrating her research on the demarcation of the line of unreliable rainfall, the survey and settlement of the Northern Territory, and Goyder's role as Protector of Aborigines.[59]

---

[53] Goyder Kerr, *Surveyors*; Sheldrick, *Nature's Line*, 222-3.
[54] Sheldrick, *Nature's Line*, 2-6.
[55] ibid, 29.
[56] Williams, "Practical Geographer", 17.
[57] L. Robin and M. Smith, "Australian Environmental History: Ten Years On", Environment and History 14 (2008): 135–143.
[58] Sheldrick, *Nature's Line*.
[59] ibid.

One of remaining seven Sheldrick named was Goyder's leading role in founding forestry.[60]

Learning to learn wholes

When Goyder's leadership of forestry is interpreted within the framework of his foundational influences, it reveals the reasons for the development of his broad-based environmental literacy. Goyder's real leadership of innovation lay not in the invention of the ideas of forestry but in shepherding the multiplicity of contradictory ideas and forces that surrounded it towards a satisfactory resolution. He filtered the external system of settlement demands through his internal system of head, heart and hands. Importantly, he also appreciated the fact that others did the same, though often with quite different results.

David Goyder was not constrained in public expression in the same way his son was as chief public servant. In fact, he was a prolific writer and commentator on his life, beliefs and thought as a Swedenborgian minister, Pestalozzian school principal, doctor and pharmacist.[61] His wife Sarah shared her husband's faith and was also a trained Pestalozzian teacher. Together, they were directly responsible for George's early education until the age of eight and, as parents, guided his more traditional secondary education in Glasgow.[62] They combined a shared philosophical view in integrated professional and personal lives. They lived and taught their family and others by the educational principles of Johann Pestalozzi and the writings of Emmanuel Swedenborg.

Goyder's ability to synthesise the scientific and the social in a specific context can be seen as a direct legacy of his father's life. David Goyder was born in London in 1796.[63] His parents were well-off and his life was, at first, very comfortable. But he lost this comfort, and with it his social position, with the death of his father at the age of nine. His subsequent struggles had a Dickensian trajectory that included time spent in an orphanage and in brutal child labour. His life improved only when, as a teenager, he managed to escape an abusive apprenticeship and obtain a position as an errand boy with the aid of his cousin. Later, while living with older brother Thomas, a Swedenborgian minister, David was gradually persuaded of the spiritual and social wisdom of the philosophy

[60] ibid.
[61] Goyder, *My Battle*.
[62] Sheldrick, *Nature's Line*.
[63] Goyder, *My Battle*, 1.

and was eventually ordained too. He and his bride Sarah trained in the equally new method of Pestalozzian education. They established and ran a small school. In time, David became prominent in both the Church and Pestalozzian education, travelling Britain to preach and to assist in the establishment of schools. The large, young family of ten moved regularly until George was eight, when they settled in Glasgow.[64]

Driven by a combination of practical need and intellectual interest, David Goyder also trained as a physician and pharmacist over the course of his professional life. Sarah also became an accomplished pharmacist. Goyder's parents modelled an unconventional life that contained elements of what we might now call social entrepreneurship. They integrated spiritual insight and practical wisdom for living. Their religion was exercised as a lived spirituality, and a novel pedagogy that put the respect of the child at its centre. The integrated nature of their professional and personal lives challenged an absolute separation of public and private lives, consolidated by the Industrial Revolution, where men operated in the public sphere and women in the domestic. Goyder received from his parents an educational experience that combined intellectual, practical, personal and spiritual training. Each of these capacities can also be seen as a legacy of the powerful influence of Swedenborg and Pestalozzi on his parents.

This legacy was directly informed by David and Sarah Goyder's shared belief in the need for social reform, a feature of early nineteenth century Britain. David Williams, writing in 1975 on the foundation of South Australia, speculated about the relationship between the philosophy of the colony's planning and nineteenth-century social reformers.[65] Amongst the reformers he names is Robert Owens, successful businessman turned educationalist and social planner. What has become evident since Williams wrote is the connection between Goyder's early education and Robert Owen. Goyder's father, David, visited New Lanark in his emerging role as a leader in Britain of Pestalozzian education.[66] By 1816, Robert Owen had become famous as an early industrialist turned social reformer. He sought through publications, building partnerships and lobbying the British government to influence social change to stem the social evils rapid industrialisation was causing. In this, he shared a common aim with Jeremy Bentham, another prominent reformer mentioned by Williams as influential in the philosophical planning of South Australia.[67] However,

---

[64] ibid, 270.
[65] Williams, "Parkland Towns", 68.
[66] Goyder, *My Battle*, 187.
[67] Williams, "Parkland Towns".

Owen's emphasis on the foundational benefits to civil society of an effective and humane education was distinctive.[68] David Goyder visited New Lanark sometime between 1826 and 1834. There, he met with Owen's son, Robert Dale Owen, and visited the school to observe the nature of instruction. Owen had been trained by Pestalozzi himself in Yverdun, Switzerland. Goyder left "delighted beyond measure with all I saw and heard".[69] He shared the same reformist educational philosophy and the belief that early education across all classes of society was vital to the establishment of well-integrated, cooperative communities, such as New Lanark and New Harmony.[70] This example of David Goyder's direct connection between education and the production of a humane and sustainable civic settlement reinforces a view of the reformist atmosphere into which George was born in 1826, and which may have encouraged the young adult's later immigration to the new world in 1851.

Swedenborg's influence

David and Sarah Goyder's family lived by Swedenborg's conviction that, in order "to receive the life of heaven a man must needs live in the world and engage in its business and employments, and by means of a moral and civil life there receive the spiritual life".[71] They were not isolated in this practice of a practical spirituality. In the first half of the nineteenth century, the Romantic Movement had adopted Swedenborg as an exemplar of the spiritual, even mystical, dimensions of humanity. Ralph Waldo Emerson wrote of Swedenborg as a "representative man" of his age, in the same company as Montaigne or Shakespeare in theirs.[72]

But the young Swedenborg had trained as a scientist, not a theologian. Born in 1788 into an aristocratic Swedish family, he later graduated from the University of Uppsala. His father was one its professors and his brother-in-law, Erik Benzelius Jnr, became a leading figure of the Swedish Enlightenment.[73] As a youth, he responded to the scholarly challenges of Enlightenment thinking, of scientific discovery in

---

[68] I. Donnachie, "Education in Robert Owen's New Society: The New Lanark Institute and Schools", in the *Encyclopedia of Informal Education,* (2003), www.infed.org/thinkers/et-owen.html.
[69] Goyder, *My Battle*, 189.
[70] Donnachie, "Education in".
[71] Sheldrick, *Nature's Line*, 250.
[72] R. Emerson, *Representative Men* (London: The Waverley Book Company, 1910).
[73] I. Jonsson, *Emanuel Swedenborg* (New York: Twayne Publishers, Inc., 1971), 19.

the service of rational and material outcomes. Later, these same intellectual motivations led him to investigate spiritual and mystical phenomena. Rather than leading to a renunciation of science, such interests extended and deepened his scientific knowledge. In Swedenborg's mind, scientific and spiritual understanding offered convergent paths to knowing and functioning optimally in the world. It can be seen as another version of head, heart and hands. Connecting and applying both, he maintained an active involvement in Sweden's political development, advising the Diet on practical policy matters throughout his life. He called this dynamic synthesis of life's scientific and spiritual dimensions the Doctrine of Correspondences. As one biographer put it, "Swedenborg travelled far beyond the usual field of the sciences, far from the safe path of experience and observation, into the realms of intuitive thought".[74] His body of work shows that "a speculative theorist is sometimes able to penetrate as deeply into the exact sciences as the most devoted experimentalist".[75]

It resulted in a capacity for innovative and creative thought and practical advice. Swedenborg's readiness to range across fields of human knowledge was central to his innovative insight. He was an early advocate of what we might now call interdisciplinary thinking and research. Individual experience needed verification from collective experience, he believed. In Cyriel Sigstedt's words, Swedenborg believed that "one science meets and enlarges the other, and each successive discovery throws new light upon the preceding".[76] Swedenborg's incorporation of scientific and spiritual ways of seeing his environment, over time, gave him an exceptional ability to interpret his surroundings and apply his knowledge for the public good. The combination of being steeped in the practical realities of both social policy-making and spiritual thought led, in 1788, following his death, to the establishment of the Swedenborgian, or New, Church. Its theology sought to integrate the apparent dualism of the worldly scientific and social with the spiritual. Such luminaries as Thomas Jefferson and Ralph Waldo Emerson encouraged the philosophies of both Swedenborg and Pestalozzi.[77] Swedenborg's systemic approach to knowledge-making, that included interior and exterior realities, is also evident in Pestalozzi's development of a holistic educational philosophy.

---

[74] ibid, 149.
[75] ibid.
[76] C. Sigstedt, *The Swedenborg Epic: The Life and Works of Emanuel Swedenborg* (New York: Bookman Associates, 1971), 150.
[77] Emerson, *Representative.*

Pestalozzi's influence

Born in Switzerland in 1746 to an aristocratic family, Johann Pestalozzi thought about and acted in the world in unconventional professional ways.[78] His rejection of corporal punishment, rote learning, and passive classroom instruction, coupled with an egalitarian emphasis on the education of the poor, appealed to progressive social and educational thinkers in revolutionary Europe.[79] Like Swedenborg's, his impact was particularly powerful in new democracies such as the United States.[80]

Pestalozzi had absolute faith in the child's ability from birth to respond intelligently to the large lessons of nature and society. The role of the educator was "nothing else but the art of assisting nature's game for achieving her own development".[81] Children at his model schools, where they were also housed, were kept constructively busy from the time of waking until bedtime, so as to avoid what he saw as the formation of bad habits. They had eight hours of lessons during the course of the day, in which the objective was to develop the child's individual skills for apprehension rather than to impart facts. The guiding tenet of Pestalozzi's philosophy was that the development of the individual occurs in an indivisible connection with the external environment. The laws of nature, in Kate Silber's paraphrase of Pestalozzi, "are the same for external nature and for human nature, for man was created in accordance with nature, and his powers and faculties are, within limits, in conformity with the forces of nature".[82]

Pestalozzi's method was documented in his book *How Gertrude Teaches Her Children: An Attempt to Help Mothers to Teach their Own Children*.[83] Written as a report to the Burgdorf Society of Friends of Education, it contained a comprehensive account of the author's method to enable mothers of the working poor to emulate in their homes what he was attempting in a school setting. His work was as much about education for social equity, social mobility and the empowering of the educationally

---

[78] K. Silber, *Pestalozzi: The Man and His Work* (New York: Routledge and K. Paul, 1973) 121.

[79] W. Reese, "The Origins of Progressive Education", History of Education Quarterly (2001) 41:1, 10.

[80] B. Oberg, and J. Jefferson Looney (eds), *The Papers of Thomas Jefferson: Digital Edition*, 2009, http://searchworks.stanford.edu/view/7947161.

[81] Silber, *Pestalozzi*,125-6.

[82] ibid, 150.

[83] J. Pestalozzi, *How Gertrude Teaches Her Children: An Attempt to Help Mothers to Teach their Own Children* (New York: C.W. Bardeen, 1898).

disenfranchised as it was about the individual's deep learning.

Pestalozzi introduced his method by declaring its underpinning value system. He asserts that "with regard to the religious opinions herein advanced, I conceive they require no apology. I have examined most of the creeds of Christendom, and, after a deliberate investigation, I have chosen that which I hold to be the Lord's truth, the fundamental principle of which is Love".[84] Like Swedenborg, Pestalozzi had a vision of a just, equitable and compassionate society.[85] The New Church "was a vision of human society at its best"; Swedenborgians defined their God as "in essence, love and wisdom". The Church adopted an educative role that sought to create "a mental and spiritual climate in which the disciplines of the arts and sciences can be used to develop this form of rationality".[86] The dynamic between the head and the heart was crucial to understanding the world in its natural and human dimensions, so that only those who were "interiorly affected by truths [and] who have cultivated their intellectual faculty" had the wisdom to function effectively as productive citizens.[87] Richard Gladish explains that this new education was to be based upon the foundation of Swedenborg's concept that love is the life of man. All humanistic studies, philosophy, sociology, history, psychology in its newest guises as group dynamics, even the scientific method, were to be tested and tried for their contribution to the new science of education.[88]

The implications of Pestalozzi's method for school teachers were explained by a North American contemporary of David Goyder. Milo Williams viewed the authority of Pestalozzian teachers "as a loan from the parent, in whom it naturally resides" and believed that "love should be the principle ruling the parent and also the teacher". But it was clearly not to be affect only. "Love must be guided by wisdom", the dual bases of the art of teaching. He declared that "the teacher should govern from love, according to wisdom; and he should regard his pupils as moral, and to some extent rational beings.... If these principles of moral government could be established, the school will govern itself".

Pestalozzi's method was grounded in these principles of connecting the affective, the sensory and the cognitive for the benefit of both and for society at large. He had a social vision which contextualised his educational mission. He declared that

---

[84] Goyder, *My Battle*, x.

[85] R. Gladish, *History of New Church Education Vol. 2* (Bryn Athyn: The Academy of the New Church, 1968).

[86] ibid, 103.

[87] ibid, 12.

[88] ibid, 21.

> I start from no positive form of teaching, as such, but simply ask myself :—" What would you do, if you wished to give a single child all the knowledge and practical skill he needs, so that by wise care of his best opportunities he might reach inner content ?"

His work was with the dispossessed, but he also acknowledged the spiritual disempowerment of the materially rich. He sought to educate not "merely to plaster over the evils in schools which are enervating the people of Europe, but to cure them at their root".[89] The elaboration of his pedagogy is not an encyclopedia of teaching techniques, but rather activities that seek to engage the child in deep learning for deep understanding of the physical, social and spiritual world. However, he was under no idealistic delusion that his approach would cure the want of knowledge and good in Europe. Nevertheless, he committed himself to making a contribution towards this ambitious end.

In Pestalozzi's words, this meant "subordinating all forms of instruction to those eternal laws by which the human mind is raised from physical impressions on the senses to clear ideas". Superficial memorizing and regurgitation of facts would not do. Rather than an abundance of factual information, Pestalozzi believed that deep learning came from simplifying this accepted form of knowledge to first principles. Students' learning should be guided towards their self-discovery of knowledge in their own time. This involved not only the student's mind, but all of their sensory powers. His overarching educational purpose, he said, was

> to simplify the elements of all human knowledge according to these laws, and to put them into a series of typical examples that shall result in spreading a wide knowledge of Nature, general clearness of the most important ideas in the mind, and vigorous exercises of the chief bodily powers.[90]

He believed that "from the moment a mother takes an infant on her lap, she teaches him". She enables him to make gradual meaning of his world by "bringing nearer to his sense what nature has scattered afar off over large areas and in confusion". Through her mediation, she "makes the action of receiving sense-impressions, and the knowledge derived from them, easy, pleasant, and delightful to him". Although Pestalozzi's times would have made perfectly normal a sole focus on the mother as nurturer of the child he included, unusually, the role of the father in his schema. In

[89] ibid.
[90] ibid.

fact, he saw the radical social possibilities of fathers' close involvement in their children's lives and learning. He noted that

> if, too, the heart (and disposition)of the father were also used for this purpose; and if the helping Art made it possible for him to link with the disposition and circumstances of the child all the activities he needs, in order by good management of his most important affairs to attain inner content with himself throughout his life, how easy would it be to assist in raising our race and every individual man in any position whatever, even amid the difficulties of unfavorable circumstances, and amid all the evils of unhappy times, and secure him a still, calm, peaceful life.[91]

Pestalozzi's view of the interconnectedness of the educational and social imperatives of his day continued. He drew causal connections between the loss of deep learning and the rise of its pretence through rote learning of information in books. This, he considered, had led to an excessive dependence on "word-knowledge", or theoretical knowledge. The abstractions that the printed word permitted had led to the sacrifice of the "essentials of all teaching" to the "hurly-burly of isolated teaching of special things". He believed that this had killed the spirit of real truth "by dishing up all kinds of broken truths".[92] Such instruction was unable to reduce its methods to "elementary principles" or "elementary forms" and so kept the instructed at a distance from integrated knowledge and meaning. Pestalozzi condemned such instruction, declaring it "unable by any of its unconnected methods to attain the end of all instruction – clear ideas; and even to make those limited results, at which it solely aims, absolutely certain".[93] Concentration on the printed word and knowledge transmission through it led, in his view, to Europe "sink[ing] into the error, or rather the insanity, that really underlay it".[94] On the one hand Europe "rose to a gigantic height in special arts and sciences". On the other, it simply "lost all foundations of natural teaching for the whole race" through the disconnection modern education had wrought between sense-impression and deep learning.[95]

Pestalozzi wanted to see a new authenticity in education that taught the relationship between human nature and the rest of nature. Man's learning had become alienated from its deep roots in ways of seeing and making meaning. He believed it had become fragmented and unintelligible

[91] ibid.
[92] ibid, 229.
[93] ibid, 221.
[94] ibid.
[95] ibid, 222.

except within these siloed fragments. His system sought a renaissance of learning and teaching that re-integrated the parts of the whole. Its structure looked first to "Nature" to deliver and bring into some sort of meaning the "sense-impressions" of life.[96]

He looked to the wisdom of the ages to determine what had been "put into the hands of humanity to strengthen this influence of Nature in developing intelligence, energy, and virtue". He considered the methods of speech, the arts of drawing, writing, reckoning and measuring to be the foundations of teaching and learning.[97] Children were assisted in their learning by the "Art" of teaching. This support enabled "the common basis of our mind, by means of which our understanding combines those impressions which the senses have received from Nature" to represent these "as wholes, that is, as concepts".[98] The timing of learning was also critical and ought not to be artificially rushed. It should keep pace with the maturing sensibilities of the child otherwise "overhasty work upon the human mind becomes a source of physical atrophy, which must inevitably result in one-sidedness, warped judgment, superficiality, and error".[99]

For both Swedenborg and Pestalozzi, there was an immanent core to their theology that was natural and so, for them, deeply human. The separation of spiritual and physical man and nature simply made no sense of lived experience to either man. Pestalozzi may have been seen as heretical in the eyes of many of his contemporary churchmen by suggesting the divine nature of man rather than his wretchedness in relation to an omnipotent deity. He summarized the essential beliefs on which his "Art" was founded:

> this law of your nature converges as a whole towards another. It converges towards the centre of our whole being, and we ourselves are this centre. Man! never forget it! All that you are, all you wish, all you might be, comes out of yourself. All must have a centre in your physical sense-impressions, and The Mechanism of Nature this again is yourself. In all it does, the Art really only adds this to the simple course of Nature. That which Nature puts before us, scattered and over a wide area, the Art puts together in narrower bounds and brings nearer to our five senses, by associations which facilitate the power of memory and strengthen the susceptibility of our senses; and make it easier for them by daily practice

---

[96] ibid.
[97] ibid, 317.
[98] ibid, 318.
[99] ibid.

> to present to us the objects around us in greater numbers, for a longer time and in a more precise way.[100]

David and Sarah Goyder modelled for their children of seeing and making meaning of the world, including the courage to think about and see it whole for themselves. George Goyder's early education came directly from his parents, unmediated by other institutional influences. The first principles of Swedenborg and Pestalozzi became, for him, hard-wired. In his parents, Goyder could observe the value of continuous learning and the practical application of the knowledge acquired to life. His early education was both humanistic and scientific, but it was his ability to make sense of the brand new knowledge of his southern hemispheric home that equipped him to act decisively in the unchartered world of a brand new colony. He had learnt the lessons of head, heart and hand well.

Remembering wisdom: Pestalozzi from the 19th to the 21st century

A later pupil of the Pestalozzian method was scientist and reformer Albert Einstein.[101] He was a recalcitrant high-school dropout before being enrolled at age sixteen with his sister at a Pestalozzi school in Aarau, Switzerland.[102] He made the observation above about this school. He responded powerfully to Pestalozzi's first principle of education that the "inner individuality" of each child be nurtured as the focus of learning.[103] The responsibility to reach their own conclusions, coupled with staged steps in the learning process, including hands-on observations, intuitions, conceptual thinking and visual imagery proved educationally liberating for the young Einstein in 1895.[104] His enthusiasm for the Pestalozzian method was not simply intellectual; he flourished in a learning environment that integrated the heart and hand with the head. He compared this period of his education with an earlier one, saying:

> when compared to six years' schooling at a German authoritarian gymnasium it made me clearly realize how much superior an education

---

[100] ibid, 321.

[101] D. Isaacson, *Einstein: His Life and Universe* (London: Simon and Schuster, 2007), 26.

[102] ibid.

[103] Pestalozzi, *How Gertrude*, 334.

[104] Isaacson*, Einstein*, 26.

> based on free action and personal responsibility is to one relying on outward authority.

The education that genuinely taught Einstein, as well as Goyder, how to learn is being revived in Europe for the twenty-first century.

Pestalozzi and his method have been articulated anew in the Council of Europe's Programme for Teacher Education which carries his name.[105] Liutauras Degesys reaffirms historians Appleby, Hunt and Jacob's description of the cultural privileging of science in contemporary teaching and learning. He suggests that if truth were absolute, it could be codified and communicated through transmission from one generation to the next.[106] New truths could be researched by scholars as derivatives of the original creed. If the world were entirely intelligible through a material or mechanistic lens, then the education of the young would be definable within these parameters. The power of man to know and therefore control external reality had been established over almost four centuries. Learning, said Degesys, is based on the premise that

> if the world contains a finite and definite number of things, if the world itself is a comprehensible "thing", then the knowledge one can gain about such a world is also finite ("definite"); one can understand and know the world in a finite way.[107]

This is an apt description of Finniss' and Brown's beliefs about knowledge of forestry.

Pestalozzi himself had lamented that such a construction of learning contradicted not only external reality but the internal reality of the learner. Even at the end of the eighteenth century, he was saying, that "in the existing methods of popular instruction these laws (of human and physical nature which allow both to flourish) are not only ignored, but generally rudely opposed".[108] The "Art" of instruction should instead guide, but also be guided by, the child, enabling students to construct their own understandings of the world in authentic ways, meaningful to them.

---

105 J. Huber and P. Mompoint-Gaillard (eds), *Teacher Education for Change: The Theory Behind the Council of Europe Pestalozzi Programme*, (Strasbourg: Council of Europe, 2011).

106 L. Degesys, "Education as Liberation of the Self: Principles and Concepts of Learning and Teaching in the Pestalozzi Programme", In J. Huber and P. Mompoint-Gaillard (eds), *Teacher Education for Change: The Theory Behind the Council of Europe Pestalozzi Programme* (Strasbourg: Council of Europe, 2011).

107 ibid.

108 Pestalozzi, *How Gertrude*, 323.

What was required of teachers was that they focus for the child "that which Nature puts before us, scattered and over a wide area" so that students are exposed in a meaningful way to the lessons nature readily offers. That focus ought to be sensual as well as intellectual so that a genuinely integrated knowledge is formed by the child. The "Art" of the teacher, Pestalozzi believed, is to frame what nature has to offer and bring that "nearer to our five senses, by associations which facilitate the power of memory and strengthen the susceptibility of our senses; and make it easier for them [the children] by daily practice to present to us the objects around us in greater numbers, for a longer time and in a more precise way".[109] The essential role of the teacher is to enable the child to come to their own understandings in their own time.

Pestalozzi named the key principles of his practice.[110] In summary, the messages they contain are the need for deep understanding and self-teaching that begins with experience which is as practical as possible; the backgrounding of teaching and teacher to the subject or object of learning (the problem or phenomenon, as well the child); the role of teaching in planning and foregrounding the latter; and doing this in staged ways that do not overrun the child's capacity to understand, and in ways that do not detach the subject or object of learning from its authentic broader context any more than is necessary. Finally, the wonder and joy of the natural world and its lessons ought to be imitated in the "Art" of the child's learning and the teacher's guidance toward that end. Education should never be dull – and it would not be, declared Pestalozzi, if it stayed close to the natural. By such means "the richness of its ("Art" imitating "Nature") charm and the variety" would inspire learning that was essential to the child's ability to function successfully in the world but do so in the guise of "free play [that tries] to make its results bear the impress of freedom and independence".[111] As Degesys noted, the object of learning was not to be made a "thing" through its detachment from its natural context or its subjective relationship to the student.[112] Only if both were fully intelligible to the teacher could they hope to impart the vital importance of the rational, emotional, spiritual and/or practical lessons to be learnt. The results of learning and teaching, said Pestalozzi, depended on "the proportion of nearness or distance between the object and our

---

[109] ibid.

[110] ibid, 320.

[111] ibid.

[112] Degesys, "Education as liberation", 47.

senses" which always has "an immense effect in determining your positive sense-impressions, practical ability, and even virtue".[113]

Contemporary commentators argue the striking difference between these essential Pestalozzian values and what came to be valued in "classical" education. Degesys explains the operating assumption of the latter as the view that sufficient transmission of disconnected knowledge would enable learners eventually to apprehend the world around them, whether this was social, environmental, political, economic or internal reality.[114] Instead, he said, the true liberation of the self, made possible through education, came from the intersection of the system of the learner with the system of what was being learnt. This idea resonates with the idea of "wicked" problems being soluble only through "systems of systems". Other commentators describe this process of learning as the "convergence of competences" or the integration of specialist or subject-specific knowledge through "transversal knowledge, skills and attitudes".[115] The holistic pedagogy of Pestalozzi - the engagement of head, heart and hand - is described by Pascale Mompoint-Gaillard as three "saviors-savoir, savoir-faire, savoir-etre". [116] In order to be a capable teacher as well as learner, she argues, an interrogation of and settling on one's authenticity and values-base is unavoidable. Likewise, the authenticity of connecting "Nature" and the "Art" of learning and teaching depends in the first instance upon this interconnection being established within the teacher.

The Council of Europe's adoption of Pestalozzi's name for its twenty-first-century teacher education programme reflects the same vision of sustainable democratic societies in an education system which aims to prepare future generations to be engaged citizens. Modelling of the various competences is the essential element of instruction, replacing the canonised version of didactic transmission of "knowledge for life" in classical education.[117] The learning environment should provide an institutional model of the society for which active citizens are being prepared.[118] At critical times in human history - such as war, revolution, catastrophe and settlement - there is the unavoidable recognition that societies and their structures are not static entities, in the same way that knowledge is not a static "thing". Pestalozzi formed his practice and philosophy in the midst of such a time, and the twenty-first century has had its own reminders,

---

[113] Pestalozzi, *How Gertrude*, 320.
[114] Degesys, "Education as Liberation", 47.
[115] Huber et al., *Teacher Education*, 1.1.
[116] ibid, 14.
[117] ibid, 19.
[118] ibid, 18.

perhaps most notably the borderless "war on terrorism" or the environmental threat of climate change. The tragedy of Black Saturday was a local example of our learned disempowerment in the face of such uncertainties. At such times, the Pestalozzi Programme proposes that

> education, which aims at the formation of active citizens, has to supply learners with the competence to learn and to constantly renew skills of accessing and handling information as well as knowledge stocks. It has to teach learners the art of learning – and it has to do it in a way which makes learning a joyful personal enterprise.[119]

Learners need to develop a self-reflexivity that makes them "not only objects but also subjects of their own learning experience".[120] Because there is no absolute canon of knowledge, learners become more dependent on their own skills in ways of thinking about their world. They need to be "given instruments for investigating and negotiating knowledge".[121] This is the "practical realism" of personal management and integration of knowledge of Appleby and her co-authors, a valuing of objectivity and subjectivity integrated in ways that are rigorous and balanced.[122]

Such a process in no way seeks to dilute the importance of subject or discipline-specific knowledge or epistemologies, but rather to infuse them with "transversal skills" and capacities that aim to deepen and broaden individual and collective understanding.[123] The classical idea of education, by contrast, was an "objective process of objective knowledge transference" that functioned in a clinical machine-like manner, transmitting a "material collection of knowledge pieces into the no less material head of the learner" who was assumed to be ready to receive them.[124] In critical cultural and social settings, "most processes cannot be simply foreseen, planned and rationalized".[125] The core principles of Pestalozzi's teaching and learning regime prepared students to manage in an ambiguous world - the kind that characterised the life for which the Goyders were preparing their children.

---

[119] ibid.
[120] ibid, 20.
[121] ibid, 21.
[122] ibid, 22.
[123] ibid.
[124] Degesys, "Education as Liberation", 53.
[125] ibid.

Pestalozzi himself was strongly influenced by Jean Jacques Rousseau. Rousseau warned against the imposition on the learner of an established body of knowledge as weighty and dead fact. Instead, said Rousseau,

> if you call your pupil's attention to natural phenomena you will soon make him curious; but in order to feed his curiosity, never be in a hurry to satisfy it. Place questions within his reach, and leave him to resolve them. Let him not know anything because you have told him but because he has grasped it for himself; let him not learn science, but invent it. If you ever replace reason by authority in his mind, he will no longer reason; he will become nothing but a plaything for others' opinions.[126]

The concluding section will say more about the linkages between Goyder's education and twenty-first century education for environmental leadership, and anticipate the same of Pinchot's.

Conclusion

Finniss and Brown's educations had, in quite different ways, been grounded in the absolute truths of transmitted knowledge. In Finniss' case, this was overlain and reinforced by the militaristic command-and-control relationship he had with his family, and with his father in particular. The English public school system, with its classical curriculum and hierarchical authority structure, both between boys and masters and senior and junior boys, reflected and consolidated a worldview in its students. For Finniss, the uncritical authority his father expected as military professional pervaded his education and his school and subsequent professional relationships. They were defined by who or what had the superior claim to authority in knowledge or in organisational structure.

Brown's educational and filial experience was more benign and benevolent, but it was encased in the singularity of forestry and his father's extraordinary efforts to elevate and professionalise the field as authoritative science. Finniss' strength and drive of youthful professional self-confidence was invested with his father's. This worked powerfully on behalf of forestry's innovation in Australia. His public promotional and educative work may have been based on faulty science and an overweening

[126] D. Leclercq, "The Pedagogical Foundations of the Pestalozzi Programme", In J. Huber and P. Mompoint-Gaillard (eds), *Teacher Education for Change: The Theory Behind the Council of Europe Pestalozzi Programme* (Strasbourg: Council of Europe, 2011), 73.

belief in the redemptive power of forestry, but it raised the profile of the work nevertheless.

Goyder's early education and the familial context in which it was received combined and integrated external systems of knowledge with internal. It not only developed an ability to think across physical, social, and humanistic boundaries, but also encouraged a valuing of knowledge, or wisdom, grounded in an internal system that combined rational, emotional and experiential learning.

The nineteenth-century educational histories reconstructed in this chapter demonstrate that the interrogation of the particularities of lives and contexts during the Industrial Revolution both confirm and challenge Deming's and Senge's conclusions about schooling during this era. This is confirmed by the revival of the principles of Pestalozzi's method for twenty-first century education by the European Council. The significance of the challenge for systems thinking is the demonstration of the constructive contribution that the historical method can make to suggestions and arguments for reform. The significance of the historical research for the work of the European Council is that it can provide models and case studies from the past that can render more concrete and exemplify the abstractions of the descriptive language of their theoretical explanations.

The educational histories of each of these leading figures in the innovation of Australian forestry offer positive and negative models of education for environmental literacy for children and leaders that have application today. They also provide substantive evidence that confirms the educational and systems thinking literature about the necessary attributes of citizens and managers to manage the uncertainties of complex environmental problem-solving in the twenty-first century.

However, the particularities of Goyder's positive learning for environmental leadership also make generalisation from this single case study problematic. Michael Woolcock, Simon Szreter and Vijayendra Rao, describing the role history can play in contemporary evidence finding and problem solving, say that

> even as most historians share with social scientists [and scientists] a commitment to generating and testing hypotheses (that is, to inductive and deductive reasoning), and recognise that the veracity of a given explanation is stronger the larger the number of cases it can explain, the canonical skill of historians is being able to immerse themselves

> sufficiently in the full context of a period or a juncture faced by those in the past.[127]

This tension between the value of history revealing the complexities and contingencies of context, and the interest in findings that can be generalised and confidently applied to a larger population, is shared by researchers of leadership. David Day, John Fleenor, Leanne Atwater, Rachel Sturm, and Rob McKee recently conducted a meta-analysis of articles on leadership over the past twenty-five years in *The Leadership Quarterly.* Their purpose was to arrive at recommendations for future research on leadership development. They argued the importance of the traditional focus on individual leaders being extended to include research on collective leadership attributes.[128] It is beyond the scope of this book to construct multiple historical case studies of learning for leadership, but the educational history of another recognized environmental leader whose context is politically, socially, economically and environmentally different from Goyder's adds weight to the preliminary findings that can be drawn from the comparative case study of leadership presented in the preceding chapters.

The following chapter constructs the educational history of Gifford Pinchot, the acknowledged and celebrated "father" of forestry in the United States. It does so in order to make a more robust argument for the value of history to management theory, and to contribute new historical knowledge to the scholarship on Pinchot.

---

[127] M. Woolcock, S. Szreter and V. Rao, "How and Why Does History Matter for Development Policy?", Journal of Development Studies 47(1) (2011): 78.

[128] D. Day, J. Fleenor, L. Atwater, R. Sturm, and R. McKee, "Advances in leader and leadership development: A review of 25 years of research and theory", Leadership Quarterly 25 (2014): 79.

CHAPTER 10

GENERALISING THE LESSONS OF SUCCESSFUL LEADERSHIP: THE USA

> The historian is not really interested in the unique, but in what is general in the unique.
> —E.H. Carr[1]

The complexity of innovating forestry was not confined to Australia in the nineteenth century. As is true today, the nature of environmental sustainability then had global reach. As did the need for leading social change in policy and behaviour. So far, the historical narrative has reconstructed the nature of the Australian innovation and its thought and action leadership. It has compared the approaches of Finniss, Brown and Goyder and examined the causal connection between the youthful experience of each man and their different adult capacities. It has also demonstrated the origins of Goyder's successful stewardship of environmental innovation over a protracted course towards a sustainable end point. However, the particularity of the Australian story, while suggestive of leadership lessons, is limited. Historians generally, and management historians specifically, acknowledge the public value of their discipline, both in searching out the unique and of aiming for generalisation.

The history of the innovation of formal Australian forestry has offered a grounded case study of what "wicked" problems actually look like, while demonstrating their precedents in the past. It has also, in the words of Woolcock et al., offered a "sobering realisation of the kind of time-scale" that current leaders of such problems should envisage for their policies and practices to come to fruition.[2] The comparative educational histories in turn provide an insight into successful and unsuccessful leadership to suggest core principles for such education in the present,

---

[1] Carr, *What is History?*, 63.
[2] Woolcock et al., *How and Why Does History Matter*, 84.

which can be overlooked in the large, negative generalisations about historical epochs such as the industrial age. The value of such examples to current management research and practice is in the distillation of essential educational principles that may be applicable across the continuum from childhood to adult learning for contemporary leadership. In order to strengthen the argument for the educational principles evident in Goyder's learning, this chapter looks at the learning history of another significant environmental leader. Pinchot's circumstances were different in time, place, scale and social maturity to Goyder's so distilling what is similar in their learning environments strengthens the argument for essential principles.

It is beyond the scope of the book to reconstruct the narrative of the innovation of forestry in the United States, and this has been well documented by historians elsewhere.[3] Its different complexity is assumed. Pinchot was instrumental in establishing the first United States Forest Service in 1905 under the presidency of Theodore Roosevelt, after three and a half decades of debates and action on behalf of forestry. A focus on his educational history aims to counter the methodological problem of leadership biography identified by Jones and Wadhwani: that "the primary drawback with such studies arises from deriving meaningful generalizations about entrepreneurship from individual cases".[4] By examining the educational influences of Gifford Pinchot and comparing these with those of George Goyder, a more robust argument can be made for the essential learning principles that shape the judgement of successful leaders.

Pinchot's educational history is reconstructed and framed by the extensive biographical literature on the man, particularly the rich and rigorous historical research undertaken by Char Miller. The chapter builds on the existing research conducted by Miller on Pinchot's early education by examining part of the comprehensive archive donated by the Pinchot family to the Library of Congress in Washington DC. It offers new historical insights to the understanding of Pinchot's leadership on behalf of forestry by drawing connections between the evidence of his early learning environment and his adult leadership ability. It also draws comparisons between this evidence and the educational history of George Goyder.

There were significant differences in Goyder and Pinchot's external circumstances. Goyder was supporting the settlement of a very

---

[3] For example, H. Steen, *The Conservation Diaries of Gifford Pinchot.* (Durham: Forest History Society, 2001); Miller, *Gifford Pinchot;* J. Lewis, "The Pinchot Family and the Battle to Establish American Forestry", Pennsylvania History 66(2) (1999): 143-65.

[4] Jones and Wadhwani, "Entrepreneurship".

new and small British colony in the southern hemisphere, before the federation of the separate colonies into a single nation in 1901. Pinchot was operating at a national level in the republic of the United States of America, which had forged its independence from Britain in the late eighteenth century, over one hundred and fifty years after Europeans began to settle the continent. The scale and timing of each man's work differed. The US Forest Service was established three decades after the South Australian Forest Board.

Pinchot's personal circumstances also bore little resemblance to Goyder's. Gifford had been born into a family of considerable wealth, commercial acumen and social standing. His early education took place in the USA, France and England enveloped by the family's more than comfortable circumstances. His father was a successful private entrepreneur who had forged a thriving business in fashionable New York in interior design.

However, Pinchot shared with Goyder an education that enabled a holistic worldview. While this was not framed by a deliberately adopted educational method like Goyder's, it laid the basis of the same capacity to construct knowledge and apply it to environmental innovation. The summary phrase of Pestalozzian educational philosophy of "head, heart and hand" was evident in the foundational principles of Pinchot's education. This chapter argues that Gifford Pinchot, like George Goyder, received an education deliberately designed by his parents to foster an internal system of humanist values and an external system able to think across boundaries of scientific, economic, social and political. And, while the USA was not undergoing the same uncertainties of early settlement, it was at the time of Gifford's birth and youth, in the throes of national rebuilding following the devastation of Civil War. Gifford's education was strongly influenced by his parents' powerful reflective responses to what may be conceived of as the re-settlement of the American nation.

An understanding of Goyder's powerful influences will also be shown to complement the extensive archival record of Pinchot. Each man's historical record offers the opportunity to cross-fertilise the other's story, sometimes directly through the evidence and sometimes inferentially. Pinchot came from a prosperous New York family in which letter and diary writing were an expected part of daily life.[5] As a consequence, the family produced a wealth of primary source material on which to draw. Rich secondary sources have also been written on Pinchot's personal and

---

[5] *Gifford Pinchot Papers: A Finding Aid to the Collection in the Library of Congress*, Manuscript Division, Library of Congress, Washington, D.C., 2011.

professional life. On the other hand, the philosophy and method of Goyder's education has been made highly accessible through the writings of his father, and the primary and secondary sources available for Pestalozzi and Swedenborg. Such rich evidence is less available for Pinchot, but there is evidence which suggests similar educational influences, the impact of which may be appreciated through what we know of Goyder's background. By pointing to the similarities and differences of Goyder's and Pinchot's education, an international story of leadership in environmental literacy for the wicked problem of forestry can be told. Each story can be rendered with more clarity by comparison with the other. What emerges from the vastly different, almost inverted, circumstances of Goyder's and Pinchot's adult, professional lives, is a narrative of similarity in the defining first principles of how they learned to learn and act in their worlds through as environmental innovators through the enabling of their early education. Their learning not only enabled the synthesis of often binary views about the place of man and nature, but of objective comprehension of fact and subjective appreciation of values and emotions. At a foundational level, the first principles of their learning to learn suggest strategies for teaching today's citizens and leaders about how to approach environmental wicked problems.

Introducing Gifford Pinchot

Ken Burns' documentary "The National Parks: America's Best Idea" provides a visual history of the practical and philosophical foundation of national parks in the United States. In the section of the film which deals with the period from 1890-1915, John Muir's and Gifford Pinchot's opposing views of the role of forests are described. Pinchot is portrayed as a hard-nosed utilitarian conservation advocate. Muir, by contrast, is seen as champion of the spiritual and aesthetic value of the natural world. Preservation in its pristine state is core to Muir's advocacy. His passion was to ensure the access of future generations to the deeply restorative powers nature has to offer as counterpoint to the dehumanising elements of the world of commerce and industry. While the conflict is portrayed as binary, with Pinchot on the side of the money makers, the documentary also offers a corrective piece of commentary to this popular but reductionist view.[6] In an interview in the documentary, the environmental historian William Cronon suggests that, rather than polarized opposites,

---

[6] Burns, Ken, "The National Parks: America's Best Idea." (2009). https://www.pbs.org/kenburns/the-national-parks/.

each man represents one of "two sides of the same coin".[7] Char Miller, Pinchot's biographer, elaborates this more complex interpretation by reconstructing the story of the controversial Hetch Hetchy Dam.

Miller describes the men's shared love of the natural world. But at a specific time in history, a vital decision became necessary between two competing goods: the subsistence of man and the preservation of the beauty of the natural valley. The choice for Pinchot was clear. Muir vehemently disagreed. If it was not possible to provide both a sustainable water supply to the citizens of San Francisco and keep the truly awe-inspiring valley, then the choice had to be for the people.[8] "The greatest good of the greatest number…in the long run" is the famous phrase of first principle espoused by Pinchot on how such judgements ought to be made.[9] The phrase could be equally attributed to the more muted voice of George Goyder. The essence of the event mirrors the one that inspired Goyder's declaration, through the use of Humboldt's voice, of the first principles on which he formally objected to Brown's first annual report on forestry in 1879.[10] When it boiled down to a fundamental choice between the competing goods of forest conservation and human subsistence, the latter for both men had to take precedence.

The education of Gifford Pinchot extends Cronon's metaphor, in that it demonstrates that, by the time he became the father of forestry in the United States, he had in fact learned to see, think and act on both sides of that coin – though not always as publicly as Muir. He acknowledged the inevitable dynamic, in the developing industrial democracy of the United States, between man and nature, and actively sought ways to embrace and manage that dynamic both within himself and the political and social world in which he was a leading public servant. This chapter will argue that Pinchot, like Goyder, learnt how to do so against the background of the social uncertainties of reconstituting post-Civil-War American society. Foundational to this ability was an immersion in his parents' lives and values, and their translation of these into the education given him as a young boy. The history of the young Gifford's learning to learn argues that he became imbued with an ability to see and experience the often conflicting parts of the whole. That Pinchot's environmental innovation happened a generation after Goyder's, on the opposite side of the world,

---

[7] W. Cronon, Episode 2, "The Last Refuge", in "The National Parks: America's Best Idea" (2009).

[8] Miller, *Gifford Pinchot*, 138-141.

[9] "Pinchot and Utilitarianism", https://www.fs.usda.gov/greatestgood/press/mediakit/facts/pinchot.shtml.

[10] *SAPP*, no. 83, 1879.

on a continent much more fecund, populated and matured as a "new world" settlement, and from the comfort of wealth and influence, does not detract from the similarities of the first principles by which each of these social entrepreneurs learned to lead.

By the time the US Forest Service was launched in 1905, Pinchot had become a powerful advocate for sustainability in a nation led by the most influential and progressive President of the United States in environmental matters (at that time), Theodore Roosevelt. In his autobiography, Roosevelt acknowledged the key role played by Pinchot in his administration:

> Gifford Pinchot is the man to whom the nation owes most for what has been accomplished as regards the preservation of the natural resources of our country. He led, and indeed during its most vital period embodied, the fight for the preservation through use of our forests .... He was the foremost leader in the great struggle to coordinate all our social and governmental forces in the effort to secure the adoption of a rational and far-seeing policy for securing the conservation of all our national resources. ... I believe it is but just to say that among the many, many public officials who under my administration rendered literally invaluable service to the people of the United States, he, on the whole, stood first.[11]

Pinchot's family

Pinchot's family had made their money in commerce. On his father's side, he was born into the second generation of a family of French political refugees with strong commercial proclivities. Miller describes how Gifford's great grandfather, Constantine, great grandmother, Maria, and grandfather, Cyrille, had fled France for America in 1816. They were supporters of Napoleon, and amongst those persecuted as part of the White Terror following the restoration of the Bourbon monarchy and the Emperor's exile.[12] Refugees, perhaps, but not poor. Constantine had run a business in Breteuil, and had managed not only to sell this before fleeing France, but also to transport sufficient assets with him with which to start a business in New York. This thrived, and the family moved to the more familiar context of the countryside. By 1819, they had purchased 400 acres of farmland and a town lot for a store in Milford, Pennsylvania.[13]

[11] T. Roosevelt, *An Autobiography* (New York: Macmillan, 1913).
[12] C. Miller, "All in the Family: The Pinchots of Milford", Pennsylvania History 66: 2 (1999): 118.
[13] ibid, 119.

Together, the father and son exercised considerable entrepreneurial intelligence, increasing the family's wealth considerably in the 1820s. Cyrille demonstrated a particular astuteness in land speculation. He acquired extensive holdings throughout north eastern Pennsylvania and New York State, and later in Michigan and Wisconsin. These included richly forested lands from which a thriving logging industry soon sprang. However, by the 1840s, Milford's prosperity and prospects had been substantially altered by the advent of rail. The town was bypassed. Although Cyrille continued to be Milford's richest tax payer, by the 1850s his older children, including second son James, Gifford's father, had to leave town to make their own way. They chose to do so in New York. Here, James not only thrived materially through the establishment of his own commercial pursuits, but did so incorporating the cultural interests of his new cosmopolitan home. Their father's and grandfather's rural prosperity had been the platform for their urban and more sophisticated lives, but James continued to incorporate his love of nature and the landscape in his new ventures, ultimately embodying all of these elements in the construction of the family's first home, "Grey Towers", a French Breton design built in Milford in the mid-1880s.[14]

James combined an aesthetic sensibility, commercial acumen and unrelenting industry to create a highly successful interior furnishing business.[15] He manufactured in New York and Pennsylvania and imported from England and Europe, and in so doing reconnected the family with its French origins, and political and cultural inheritance, through art and commerce. James' patronage of the arts and artists was consistent with his wealth, urban cultural context and aesthetic interests. He was drawn particularly to the late generation of the Hudson River School. A favourite was Sanford Gifford. He became Gifford's namesake and godfather at the boy's birth in 1865, a year after James married Mary Eno.[16] These happy events were no doubt an antidote to the national grief following the end of the Civil War and the assassination of a beloved President in the same year. The American Civil War of 1861-65 remains the bloodiest conflict in American history. More than six hundred thousand men perished.[17] Sanford Gifford was an artist who had served as a soldier during the War.

---

[14] Steen, *Diaries*, 51.

[15] N. Pittman, "James Wallace Pinchot (1831 – 1908)", Yale Forestry and Environmental Studies News, Fall (1999): 4-7.

[16] Miller, *Gifford Pinchot*, 32.

[17] I. Finseth, "The Civil War Dead: Realism and the Problem of Anonymity", American Literary History 25(3) (2013): 535-562.

He embodied the personal as well as civic need for national reconciliation into which his godson had been born.[18]

The Hudson River School shared its philosophical foundations with those of Ralph Waldo Emerson and Henry Thoreau. Its subject was landscape painting, its philosophy Romanticism. The combination of the School and the writings of Emerson and Thoreau was prominent in the preservation movement. Max Oelschlaeger argues that the contemporary environmentalist movement owes a lot to these artists. The Hudson River School produced "canonical landscapes". Emerson was instrumental in encouraging the national quest for identity, and Thoreau was the nation's most original philosopher of nature.[19] James Pinchot's independent fortune enabled a growing patronage of the School, as well as his dissemination of their works through travelling exhibitions in galleries nationally and internationally.

Arguably, however, the most poignant of Sanford Gifford's works remained in James', and, subsequently, Gifford's continued possession.[20] "Hunter Mountain, Twilight" was painted in 1866, a year after the War's end. The work is a melancholic, yet strikingly beautiful, depiction of a ravaged landscape. It describes a scene of extensive damage done to nature through man's violent domination. But its poignant use of light suggests that the beauty and hope of the scene cannot be expunged forever. The sad hope of restoration is conveyed through the soft, mellow tones of a twilight setting.

Such hopeful paradox resonated in James' and Mary's sensibilities. The impact of a catastrophic war sat alongside a joyous marriage and birth of a first son, followed closely by the young family's regular and extended visits to France. Such times of happy and sad reflection may have returned James to a consideration of the roots of his father's and grandfather's French republican idealism, seen through the eyes of his own American upbringing and adult experience of war. Abraham Lincoln's powerful republican ideals formed the philosophical context of this early period of James' and Mary's family life, into which Gifford was born. Their intimate friendship with General William Sherman, kept active by a

---

[18] S. Richardson, "A Day in the Life of a Soldier: The Illuminating Civil War Paintings of Hudson River School Artist Sanford Gifford", American History 49(1) (2014): 44.

[19] M.Oelschlaeger, "The Roots of Preservation: Emerson, Thoreau, and the Hudson River School", http://nationalhumanitiescenter.org/tserve/nattrans/ntwilderness/essays/preserva.html.

[20] C. Miller, "Char Miller on Gifford Pinchot, Photographer", Environmental History 8:2 (2003): 312-315.

regular correspondence between the General and Mary, testifies to the sustained impact of this tumultuous period of American history on their lives.[21] Mary's health remained fragile and of concern to James throughout the '60s and '70s, and was the reason for their protracted stay in Paris.[22] David McCullough describes the "greater journey" of many Americans to Paris during the period from 1830 to 1900, encouraged by superior professional training, especially in medicine and the creative arts, and the health-giving powers promised by the city.[23] A combination of these factors may have encouraged the repeated return of the family in the two decades following the war. Fortunately, this led to the documentation of much of the family's life through their transatlantic correspondence.[24]

By the age of five, Gifford was already learning the family's practice of writing diary entries and letters to document their daily lives.[25] In a moving entry of January 1871 in Mary's pocket diary, she writes on one side of the page about the events of her day, leaving space on the opposite page for five-year-old Gifford to practice his (then unintelligible) letters. The diary also contains a pressed leaf, a simple, striking symbol of the young child's early interest in specimen collecting of the natural world. The leaf is likely to be a preserved souvenir picked from one of Gifford's daily outings to such places as the Tuileries Gardens or the Bois de Boulogne, accompanied by his mother, father, family visitors or nanny.[26] Mary, like Sarah Goyder, was a constant and defining parent in her children's young lives, although from the quite different context of wealth and privilege.

Mary Eno Pinchot was the daughter of the prosperous New Yorker Amos Eno and his wife Lucy (Phelps) Eno. Their marriage had merged two wealthy families who also had a strong sense of civic and political duty.[27] Amos Eno had consolidated a vast family fortune as a dry goods merchant and real estate entrepreneur, and was to set his sights on his grandson continuing these ventures.[28] As late as the 1890s, Amos remained contemptuous of Gifford's dedication to forestry. Gifford

---

[21] *Gifford Pinchot Papers*, Box 39, Diaries & letters James and Mary Pinchot 1867 – 1885.

[22] ibid.

[23] D. McCullough, *The Greater Journey: Americans in Paris* (New York: Simon and Schuster, 2013).

[24] ibid, Box 39.

[25] ibid, Box 39, Mary Pinchot's pocket diary 1871.

[26] ibid.

[27] Miller, *Gifford Pinchot*, 32.

[28] ibid, 128.

records a diary entry on July 1 1891, following a conversation with Amos Eno, that "Grandpa [was] still soured on forestry".[29] His grandfather was unrelenting in his efforts to persuade his firstborn grandson to follow in his footsteps. As late as 1894, Pinchot records a lucrative offer of employment from Amos that had followed some "long talks". He had "finally offered me $2500 a year to work for him – kind of work he refused to state", and had talked at length about "necessity for fitting myself for coming responsibilities".[30] His maternal grandfather also provided a general model of individual endeavour and entrepreneurship. While they were out fishing in Connecticut five months later, Gifford recounted how his grandfather "told me about early life & showed me where he first kept store. His example is a marvelous one in energy, concentration and ability of judgment".[31]

But Mary's and James' aspirations for their son challenged her father's commercial focus.[32] Char Miller has argued that the early retirement of James in the mid-1870s and the redirection of the family's energies into civic concerns, was a direct result of the provocation of Mary. Gifford's later attribution to his father of the title, he considers, "father of US forestry" omits the key role she played.[33] This omission may be seen as at least partly corrected by Pinchot's later acknowledgement of the critical role played by women in nurturing children's concepts of citizenship, including the importance of conserving the nation's natural resources.[34] The activism Gifford demonstrated later in life on behalf of women's suffrage - in 1914, the year Mary died and he married Cornelia Bryce, he was elected vice president of Men's Association for Woman Suffrage - may also be seen as one political manifestation of the combined force of the shared sensibilities of James and Mary on the personal and intellectual socialisation of their son.

James' retirement allowed him to focus on civic responsibilities and cultural interests, and to take more of a share in the parenting of his young family. His letters to his parents from Europe in the early 70s indicate an unabashed delight and genuine engagement with his children.[35] His and Mary's active patronage of the arts, architecture, landscape and

---

[29] Steen, *Diaries*, 11.
[30] ibid., 16
[31] ibid, 12.
[32] Miller, *Gifford Pinchot*, 34.
[33] ibid.
[34] G. Pinchot, *The Fight for Conservation* (Seattle: University of Washington Press, 1910), 133.
[35] *Gifford Pinchot Papers*, Boxes 39 -76f.

education also provided a basis of an intellectual life that was a model for their children. Gifford, or Gippie as he was known in family circles, developed a love and respect for both parents, clearly evident in the affectionate language of his youthful letters.[36]

Pinchot's education

Gifford's early years and education can be seen in the context of both post-war reconstruction and the educational reform movements of the mid nineteenth century. Such reform was represented in the United States by Romantics, such as Thoreau and Emerson, and interest in English social philosophers, such as Herbert Spencer.[37] During the nineteenth century, Western countries came to recognise the value of an educated population to their economies, but some social commentators have seen the regimented form that this education was assuming as both problematic and persistent over time.[38] Gifford, by contrast, received an education that bore little resemblance to this regimented style. In both content and pedagogy, his learning continued the encouragement of "renaissance" thought of the kind acknowledged as essential in scientific training for the development of twenty-first century skills.[39]

Gifford's schooling was not regularised until December 1881 when he entered New Hampshire's Phillips Exeter Academy.[40] Even here, the school's culture was liberal. Later in life, he reflected on its positive influence and the philosophical continuity from what he had learned to expect of education as a young boy. He said of Exeter that

> as I grow older, and as my interest in the welfare and progress of our people broadens and deepens, I look back with steadily growing satisfaction [at] the time I spent at Exeter. The most useful thing about

---

36 ibid.

37 S. Meehan, "Ecology and Imagination: Emerson, Thoreau, and the Nature of Metonymy", Criticism 55(2) (2013): 299–329; H. Spencer, *The Study of Sociology* (New York: D. Appleton, 1896).

38 Senge, "Creating Schools".

39 T. Rainsford, "Back to the Future: Educating the Renaissance Scholar", unpublished presentation to the Institute for Phototonics, University of Adelaide, 16 September 2009; G. Neale, 2010, "Wanted: a 21st-century renaissance", Oxford Today, Hilary Issue: 12-14; OECD Centre for Educational Research and Innovation (CERI), "Educating for Innovative Societies" Conference Summary Report, 26 April 2012, Paris, France; P. Griffin, B. McGaw and E. Care (eds), *Assessment and Teaching of 21st Century Skills* (New York: Springer, Dordrecht, 2012).

40 *Gifford Pinchot Papers,* Box 39, James Pinchot's letters to his parents, 1871- 75.

> the school in those days . . . was the fact that it made every student responsible to himself as well as to his teachers for what he did and how he did it, and yet gave each boy the best possible chance to follow his own bent.[41]

Miller attests to the eclectic environment that then characterised the Academy. In view of the previous sixteen years of non-traditional education that had been his norm, it is not surprising that Gifford enjoyed his time at the institution, which prized self-directed learning. It was also an environment which allowed for the considerable, if physically distant, intervention of James and Mary, who now expected much more disciplined application than their son had learned to expect of education.[42]

Miller describes the "peripatetic" lifestyle that characterised the Pinchot family during Gifford's early years and with it the young boy's educational context.[43] On the one hand, he was taught the basics of a classical education: French, dance, art, mathematics and literature. On the other, he was ensconced in a learning environment in which "the natural world was a consistent frame of reference in the family's discussions". His innate proclivities were actively encouraged and gravitated strongly toward the "biotic realm".[44]

We get a taste of this early development of a love of nature from a diary entry made when Gifford returned to Paris much later, as a Yale graduate, in search of the best path for the professional study of forestry. With letters of introduction from London to the leading German figures of European forestry, Dr William Schlich and Sir Dietrich Brandis, he went first to visit the forest exhibit at the World's Fair, held in Paris during the hundredth anniversary of the storming of the Bastille. In doing so, he was commemorating an event which had eventually led to the beginnings of his paternal family's emigration to the United States.[45] Instead, he recalled the instructive delights of his early education in Paris in a diary entry of 24 October 1889, noting "how glad I was to be in Paris again! It seemed like the good old times, almost when I was studying at the Jardin des Plantes, and the family was here".[46] Diary entries from during his time as a student of forestry in France indicate his frustration at the dry and theoretical style

---

41 Phillips Exeter Academy Bulletin,
www.exeter.edu/documents/exeter_bulletin/spring_02/newground.html.
42 Miller, *Gifford Pinchot*, 58.
43 ibid.
44 ibid.
45 Steen, *Diaries*, 32.
46 ibid.

of instruction he experienced at Nancy. When Pinchot decided to cut short a full European course of forestry study and return to the USA, a contributing factor was likely to have been the confidence generated by his early self-directed, experiential learning. He knew not only how he learned best, but also the educational principles that led to effective teaching of others about forests.

For a young boy with a powerful enthusiasm for the world of nature, the Jardin des Plantes offered a wonderful laboratory for investigation.[47] During that period of his life, he was fascinated by insects, which he would have found in abundance during daily outings.[48] Established in the seventeenth century, the Jardin exists today on the same 68-acre site. It now contains 23,500 species of plants. As the National Museum of Natural History, it remains one of the world's foremost botanical gardens. Its great expansion happened in the eighteenth century when it became a centre of scientific study for prominent zoologists and botanists. By the nineteenth century, it was supporting expeditions to distant corners of the world to acquire species then unknown to Western science.[49] Learning the lessons of nature in the man-made creation of the Jardin was one powerful source of Pinchot's introduction to the natural world, situated in the midst of human activity.

In his diary's reference to the family's trips to the Jardin, the twenty-four-year-old Pinchot was alluding to the family's extended stays in Paris in the first half of the 1870s and again in the early 80s. In 1871, the young family of four settled in Paris, with James returning when necessary for business to the United States. In doing so, they were among the many of their compatriots engaged in the transatlantic movement described by McCullough.[50] There was an established American "colony" in Paris which, before the Franco-Prussian War of 1870, had numbered close to 13,000.[51] The colony reconvened following the siege, with the Pinchots apparently being amongst the first to return. Mary's and James' third child, Amos, or Toots as he was affectionately nicknamed (after the character in Dickens' *Hard Times*), was born in France in December 1873.[52] Paris had become their home away from home, and learning the

---

[47] *Gifford Pinchot Papers,* Box 30.
[48] ibid, Box 39, James Pinchot letters to his parents, 1871-75.
[49] "Jardin des Plantes",
http://www.britannica.com/EBchecked/topic/301392/Jardin-des-Plantes.
[50] McCullough, *Greater Journey.*
[51] ibid.
[52] *Gifford Pinchot Papers,* Boxes 39, James Pinchot's letters to his parents, 1871-75.

language formed part of Gifford's early educational routine.

When in Paris with the family, James regularly wrote home to his parents in Milford of Gifford's daily morning French lessons. He explained, too, how his younger sister Nettie was also benefiting through the family conversations in French in which her brother was now an active participant. But French lessons were completed by ten in the morning, and outdoor activities until dinner time formed the largest part of the day's regular schedule.[53] Visits to the Tuileries Garden, the Bois de Boulogne or the Zoo offered fresh air and physical exercise, as well as the opportunity for investigating the world of urban nature. Visits to the Jardin in the early 70s, following the Communards' burning of the Palace, must also have offered simple lessons in social and political history.[54]

Later in the decade, when the family were again based in New York, Gifford wrote frequent letters to his teacher, Mr McMullen, about any topic which had captured his interest.[55] In one, he declared his filial as well as political connection with the French, asserting the view that the American Revolutionary War would have undoubtedly failed without the fraternal allegiance and support of the French. The letter was in English but could as easily have been in French, which was by then his second language and one he continued to practise while resident in the USA.[56] James, too, valued the opportunity to return to his father's and grandfather's home town of Breteuil, and was particularly pleased to be able to do so after having acquired a greater proficiency in the language and through it deeper insight into his French heritage.[57]

Mary's pocket diary of 1871, containing the preserved leaf and the letter-like markings of a young child, show the shared activity of adult and child in James and Mary's approach to Gifford's learning. Preserving the importance of nature and the lessons of outdoor activity while modelling the daily disciplines of recording the events of his day were to become staples of Gifford's life. By 1872, Gifford was working on his own diaries, and writing his own letters to relatives at home.[58] But his letter writing as an adolescent living away from home never became as disciplined as Mary and James wanted. Their requirement of writing twice a week on Sunday or Monday, depending on church-related commitments, and Wednesday, was excessive for a teenager who had been taught to

[53] ibid.
[54] McCullough, *Greater Journey*.
[55] *Gifford Pinchot Papers,* Box 30, Gifford Pinchot's letters to Mr. McMullen.
[56] ibid.
[57] *Gifford Pinchot Papers,* Boxes 39, James Pinchot's letters to his parents, 1882.
[58] ibid.

maintain an active and curious engagement with the world as a young boy.[59] While he wrote often of his contrition at his failure to comply with his parents' desires for disciplined communication, he nevertheless continued to exercise a physical and intellectual independence that had, after all, been encouraged in his early childhood.

Between 1876 to 1879, Gifford did write regularly to his teacher as part of his school routine in a red exercise book marked in black, "Letterbook".[60] The educational practice of a mixture of formal lessons and self-directed enquiry in the world continued through these years. When James wrote to Mary in 1882 to suggest they continue to keep nine-year-old Amos away from school for health reasons, he indicated again a parental preference for using schools as supplements to, not staples for, their children's learning. Gifford's letters which were written at the same age show an interest in insects, books, the mechanics of vehicles, hunting, and the rights and wrongs of warfare, and demonstrate the ease with which he already functioned in an adult world.[61] He was, and remained throughout his youth, an avid entomologist. By 1876, he had learned a literacy of the insect world which he was happy to describe to his teacher:

> My dear teacher,
>
> I want to tell you something about my collection of insects. This is the second year I have been at it. I have almost two hundred specimens, and when you think how many there are, this is very few, but I still have more than I thought I would have when I began. I do not confine myself strictly to these, but also take all the little curiosities I can find. I have two very nice fossils. One, a curiously shaped snail is something like this (a small, finely detailed drawing), only about five times as large….[62]

In order to communicate his point, he used whatever language was available to him. If he did not know how to describe the concentric circles of a snail's shell, then he drew them. If he could not translate from his French natural history book into English to relate the information he wanted to his American teacher, he relayed the information in French. In the same letter, he advised the teacher that he was not only depending on his capacity to collect artefacts from his surrounding neighbourhood, but was also deliberately acquiring objects of interest, in the same way that his father did: "The other day some furniture arrived for Papa. With it came

[59] ibid.

[60] *Gifford Pinchot Papers,* Box 30.

[61] *Gifford Pinchot Papers,* Box 39, 1874.

[62] *Gifford Pinchot Papers,* Box 30, 1876.

some very nice fossils for me….".[63] His interest in the natural sciences continued, and with it came a growing scientific literacy. By 1878, he spent a page relaying to his teacher the gift of a microscope and how he could manipulate it to learn more detailed information about the objects of his study. Interleaved with stories of scientific enquiry were those of the hunting and fishing of larger species, which were both objects of similar enquiry and of outdoor sports. Summaries of novels also peppered the young Gifford's letters. One retold the narrative of "The Peachling", a Japanese tale that anticipated the international flavour of the notes he would write during the family's stay Paris in the early 1880s.[64] These later notes were of lectures he attended to learn more about the animal and reptile species of the world, including the peculiar versions to be found on the island continent of Australia.[65]

Gifford's letters to his teacher also reveal how comfortable he was in, and how his education profited from, the company of his parents' celebrated friends. On February 21 1877, he set out to provide his teacher with "a short account of the first half of my excursion with General Sherman and his staff".[66] After meeting up with the General's party at the foot of 33rd street and East River, the party sailed off on the steamer "Henry Smith". First stop was Willett's Point, passing Hell Gate, where "all the soldiers on the wharf" fired a salute. General Abbott, the Commandant, proceeded to give General Sherman a tour of the facilities. Gifford relates how "I got to go with him and so got a good deal of information". Lunch was provided at the Commandant's house, "and a very good lunch it was".[67] Following the meal, Gifford boarded the steamer with Sherman and Abbott and sailed across to Fort Schuyler. At the end of his letter, Gifford suddenly remembered that he forgot to tell his teacher the story "about exploding the torpedo", but having clearly by this time satisfied the usual length requirement of letters he signs off abruptly "but I will tell you about that next time, so good bye".[68]

The young Gifford's proximity to reformers of the second half of the nineteenth century, such as Sherman, point to the family's philosophical and civic engagement with the social reconstruction of the United States following the tragedy of civil war. Immersed in both this context, and the ongoing settling of the French republic, Gifford learnt the importance of

---

63 ibid.
64 ibid, Box 30, 1878-80.
65 ibid.
66 ibid, Box 30, 1877.
67 ibid.
68 ibid.

the socio-political world and his place in it. The intimate friendship between the Sherman and Pinchot families can be measured by the frequency of correspondence between Mary and William Sherman and the Sherman family visits across continents.[69] Likewise, the significance of Mary's maternal family's role in the War itself can be measured by the collection of correspondence between John and Mary Phelps and Amos and Lucy (Phelps) Eno.[70] The closeness of the Eno's extended family, the inclusion of the Shermans in that intimate circle, and the proximity of Gifford's birth to the War's end, infused his broader education with what Sherman called in an 1863 letter to Halleck the "reconciliation and reconstruction" that would follow from the agony of the "fratricidal war".[71]

The model of heroism Sherman, John and Mary Phelps presented for Gifford was one in which they acted with a spiritual courage in an inhumane event. Stephen Bower wrote of Sherman's "theology of the battlefield", where he believed himself to be at war with the "demons of disintegration and doubt that so tormented the American soul and the angels standing guard over life, liberty and the pursuit of happiness".[72] James Pinchot's active role in the construction of the Statue of Liberty represented his contribution, through the exercise of his artistic sensibilities, to the efforts towards reconciliation and reconstruction. Gifford showed signs of his appreciation of the tragedy of the Civil War and the expectation that his family should be part of this effort at reshaping society. In a letter written three weeks before the excursion with the General, Gifford wrote to his teacher about his views on gunpowder in war. Arguing his case against he wrote,

> I do not see how gunpowder can do half so much good as it has done harm. In blasting I do not see why dynamite and nitroglycerine could not be used just as well. In war not half so many men would be killed and the victory would necessarily be won by pluck and dis[c]ipline. Just think how many lives are lost by the explosion of a single bomb shell in a body of troops, and if it were not for powder that bombshell would not be there to burst and the lives of so many men would not be lost. Think of the helpless women and children that are killed by the help of that vile invention, powder. …I leave you to judge whether it

[69] ibid, Box 39.

[70] ibid.

[71] W. Sherman, *Memoirs of General Sherman,* http://www.gutenberg.org/files/4361/4361-h/4361-h.htm.

[72] S. Bower, "The Theology of the Battlefield: William Tecumseh Sherman and the U.S. Civil War", Journal of Military History, 64: 4 (2000): 1005-1034.

does more harm than good.[73]

As well as his vicarious exposure to the Civil War through the General, Gifford understood the brave efforts of his Phelps relatives in Missouri. Although both the Eno and Phelps families were long-time residents of the north east coast, Mary's Uncle John and Aunt Mary had, since the 1830s, lived in Springfield Missouri at a time when that state represented the Wild West. The letters home that Mary Phelps kept up in the 1860s, and which Mary Eno Pinchot preserved, must have rendered even more immediate to her young niece the grave realities of the front. Her Uncle John was a Congressman who opposed secession during the crisis of 1861 and helped to keep Missouri in the Union. After the start of war, he did not stand for re-election to the Congress, but left to recruit and command a Union regiment at the Battle of Pea Ridge in March 1862. His son John also fought in the Battle under his father's command. Mary remained at the family home amidst constant fighting to nurse the wounded in her house turned hospital. But the day before Pea Ridge, she left the city for the camp, taking several wagons loaded with provisions and nursing supplies. She spent the entire battle exposed on the field attending to the wounded.[74] A letter to Amos Eno indicated that she'd been clipped in the foot by a bullet but had remained undeterred.[75] In an unusual gesture after the war's end Congress appropriated a large sum so that she could establish a home for war orphans. Years later, in 1877, John Phelps won Missouri's gubernatorial election with an exceptional majority. In a state where he needed to manage staggering war debt, inflation, depressed wages and business collapse, he persisted in championing the cause of desegregated public education. In his inaugural speech, he advocated common schools for whites and blacks and, during his term, the University of Missouri received biennial appropriations for the first time and public schools increased financial support.[76] Such were the models of citizenship that Mary Eno Pinchot brought to the family and whose stories of civic duty informed Gifford's developing sense of self.

As a young student at Exeter considering his future, he seemed aware of powerful expectations for his own future, either in commercial or

[73] *Gifford Pinchot Papers,* Box 30, 1876.

[74] "Biographical Sketch of Mrs. Mary Phelps, Dec'd, Greene County, Missouri, Springfield", from *History of Greene County, Missouri*, St. Louis, 1883. http://files.usgwarchives.net/mo/greene/bios/p4120007.txt.

[75] *Gifford Pinchot Papers,* Box 39.

[76] Missouri State Archives Finding Aid 3.23 Office of Governor John Smith Phelps, 1877-1881.

civic service. Writing to his sister Nettie towards the end of his time at the Academy, he gave expression to the stresses he had felt in struggling with the combination of poor eyesight and catching up to his peers in the academic work needed for entrance to Yale. He thought it best to admit temporary defeat and go to work for his Grandfather Eno's business for a year rather than enter Yale with the conditions he felt sure must attach to an admission, but within days he regretted his failure of emotional resilience and assured his parents that he would persist in a more "manly" way.[77] The exchange showed not simply that he felt the necessity of fortitude in the face of adversity, but also that he felt able to express the truth of his vulnerability in the first place. In doing so, he demonstrated the modelling in his life of significant men and women who contained the strength of disciplined stoicism, humane and artistic sensibilities and an expressive emotional intelligence. The convergence of these human traits and their importance to the way in which James and Mary envisaged the development of their son was embodied in their choice of godfather, Sanford Gifford, and the painting that was a visual mainstay of Gifford's life.

Some commentators believe that "Hunter Mountain, Twilight" reveals Sanford Gifford's personal story of devastation alongside his belief in the promise of hope. A review of a recent book about Sanford Gifford's work suggests that, of the Hudson River School, his paintings were quite distinctive in their "ethereal and atmospheric" qualities.[78] He admired much of J.M.W. Turners' work and sought to offer impressions that would encourage a profound response in the viewer. "Hunter Mountain, Twilight" in particular is interpreted as a representation of the temporary spiritual loss of the nation's humanity through the Civil War, symbolized by its forests.[79] As the only brother to survive the Civil War, Sanford Gifford knew its violence and violation but could not surrender the light of hope in its rendering. The same devotion that acknowledges and learns from suffering, but remains a constant source of hope, was manifest in the relationship between James and Mary.

The love between James and Mary is eloquently on display in the correspondence exchanged over the twenty years of married life from

---

[77] *Gifford Pinchot Papers*, Box 1, 1882.

[78] S. Gifford, K. Avery, and F. Kelly, (eds). *Hudson River School Visions: The Landscapes of Sanford R. Gifford* (New York: Metropolitan Museum of Art, 2003) http://www.metmuseum.org/research/metpublications/Hudson_River_School_Visions_The_Landscapes_of_Sanford_R_Gifford.

[79] "The US Civil War in Art", http://www.civilwarinart.org/items/show/93 accessed 30 November 2013.

1864 to 1884, and throughout Gifford's maturation to adulthood.[80] The spiritual and reflective basis of their emotional connection to each other, to their children, and to their own parents was also passed on to Gifford. The family's religious expression of these spiritual sensibilities was regular attendance at Christian services on a Sunday, though they were eclectic and ecumenical in their patronage.[81] Gifford absorbed their spiritual dispositions as well as their sense of choice. Both were evident in his decision to participate in a weeklong mission while boarding with the Reverend Dr Angelsmith in Canterbury in 1880.[82] Again in Exeter, he decided that the local Episcopal Church had less to offer his spiritual wellbeing than the local Protestant congregation.[83] Self-mastery and self-direction were becoming canonical as educational first principles in both his academic and personal life.

Herbert Spencer was amongst those whose visits contributed to James and Mary's reflective approach to life and their children's spiritual, intellectual and emotional upbringing.[84] James wrote to his father of how much he enjoyed Spencer's and Professor Touman's stay with them in France in October 1871. The two visitors were engaged not only in philosophical conversations with the family, but also joined them on afternoon outings with Gifford and Nettie, on one occasion to the zoological gardens. Thoreau's and Emerson's writings were also favourites of James, and both parents sent Gifford quite direct advice on what he should be reading when he was living away from them.[85]

Spencer's *Essays on Education* were written and published in prestigious journals in the 1850s and in book form in 1861. In them, he warned teachers of the intellectual and physical dangers of "excess of mental application". Supporting his views, Spencer cited the authority of Johann Pestalozzi and Scottish physician Sir John Forbes. He warned that

> if the higher faculties are early taxed by presenting an order of knowledge more complex or abstract than can be readily assimilated; or if, by excess of culture, the intellect in general is developed to a degree beyond that which is natural to its age; the abnormal advantage gained will inevitably be accompanied by some equivalent, or more than

---

[80] *Gifford Pinchot Papers,* Box 39.
[81] ibid.
[82] ibid.
[83] ibid.
[84] ibid.
[85] ibid.

equivalent, evil.[86]

He decried the imbalance in the contemporary approach to children's education. It contained too little food, too little clothing, too little exercise and an excess of mental application. It was simply cruel. As late as Gifford's entry to Exeter, James was writing him anxiously with advice to spend no more than three to four hours a day at his books, combining this with physical exercise outdoors and eating a nutritionally balanced diet, including fresh foods.[87] What had been expressed as joy in the company of his young children was now, in Gifford's developing physical independence from the family, expressed as anxious discipline. James regularly sent Gifford and his landlady at Exeter large quantities of fresh oranges and apples in season. The balance that Spencer advocated was the stuff of James and Mary's letters to their son. As independent and self-sufficient as Gifford had demonstrated himself to be from the time of his extended stay with Dr Angelsmith at age 15, he nevertheless anticipated with sheer excitement the visits of James to Exeter, and his own trips home to be with the family.[88] The mental and physical balance that Spencer advocated was matched by a balance in the emotional bond between parent and child that increasingly allowed for distance between family members. This juxtaposition of independent closeness had also been modelled for the son by the parents' relationship, in Mary's and James' long absences from one another during which they remained dedicated correspondents who openly expressed their strong bond.

Pinchot's diary entries show the balance prescribed for exercising Gifford's body, heart and mind reached into the soul.[89] In the same way that Sanford Gifford had managed to portray a scene of ravaged, yet beautiful landscape, redeemed by the radiant light in which the whole scene was bathed, the examples of literature and religious tolerance were gradually added to Gifford's repertoire of learning. His holistic understanding of the diverse and contradictory experience of life was expanded through his reading. James' recommended Emerson's and Thoreau's writing to the teenage Gifford, though with a caution that they be approached slowly and in small quantities at first.[90] These writers' philosophies contained a level of mature wisdom that James thought Gifford could only grasp fully and

---

[86] C. De Stasio, "Starving Vs Cramming: Children's Education and Upbringing in Charles Dickens and Herbert Spencer", Dickens Quarterly, 27(4) (2010): 299-306.
[87] *Gifford Pinchot Papers*. Box 39, 1881.
[88] ibid.
[89] Steen, *Diaries*.
[90] *Pinchot Papers*, Box 39, 1881.

purposefully as his experience broadened. Emerson and Thoreau dealt in both the physical and metaphysical stuff of nature. They synthesised spiritual and material divisions. For James, the writings of both contained the articulation of an intellectual, physical, and spiritual energy demanded of a wise life, in which the individual was able to continuously embrace life's complexity.[91]

His aspirations for Gifford might be seen in Andrew Taylor's description of Emerson's maintenance of a "sometimes precarious balance between scepticism and belief, keen to argue against our craving for easy certainty" while at the same time recognizing that the "abandonment of belief is a human impossibility".[92] As Thoreau's mentor, Emerson had encouraged him in journal writing as a means of "sift[ing] through… conflicting thoughts regarding the conduct of life in addition to his scholarly and professional aspirations".[93] Gifford also learnt from his parents the habit of keeping diaries that served this same variety of heuristic purposes. Through this combination of discipline and developing insight Thoreau "demonstrated a life of establishing connections, observing nature and understanding its relation to his own life and to the broader society".[94] Through the course of his life Gifford learned to do much the same. Like Emerson and Thoreau, he learnt not to see as dichotomous the transcendental and the immanent. Like Pestalozzi, Emerson had said that "the secret in education lies in respecting the student".[95]

Gifford's proclivities for learning were respected throughout his education from child to young adult. At the age of ten he chose most often to write to his teacher about insects, animals, birds and fish. When he was seventeen and at Exeter, he bought an axe so that he could fell dead trees and peel the bark in search of insects.[96] He asked his parents for his butterfly net when the season came, and began learning and recording the Latin names for the insects he was fascinated by in the notebooks he kept in Paris in 1881.[97] His surviving lecture books from Yale are those on

---

[91] ibid.

[92] A. Taylor, "A Man is Conservative after Dinner: Ralph Waldo Emerson, Michel De Montaigne and the Appetites of Moderation", Forum for Modern Language Studies, 48:1(2012): 1-18.

[93] ibid.

[94] D.Hochstetler, "Pedagogy and the Teacher Within: Reflections on Thoreau and Kinesiology in Higher Education", Quest, 65: 2 (2013): 186-200.

[95] ibid.

[96] *Pinchot Papers,* Box 1, 1882.

[97] ibid, Box 30, 1881.

botany.[98] Gifford's literary interests remained strong and ongoing but it was clear early on that, in his case, respecting the student meant encouraging his more powerful natural affinity with the "biotic realm". [99]

The Scientific Revolution of the nineteenth century, which had begun to quarantine nature as the subject of physical science alone, was insufficient in a household with a more holistic understanding of the world. Mary was concerned that, as Gifford grew to maturity, his passion for the physical and material world of nature should continue to be balanced by a parallel appreciation of the metaphysical dimensions of nature. At her instigation, the Reverend A.H. Gesner, a former minister of the Pinchot's parish at Milford, corresponded with Gifford about the integration of the physical and spiritual.[100] At a time when the scientific community was in the throes of debating Darwinian theories of natural selection and Spencer himself had coined the phrase "survival of the fittest", Mary clearly felt that she could not leave to chance the maturation of her son's spiritual sensibility. Gesner shared the young man's passionate interest in the natural world. He counselled Gifford not to become aligned with those who "persuade themselves that all the *nice*, the *fine*, the delicate adjustments and arrangements of beasts, birds, and flowers, came of themselves or were a spontaneous or developed form".[101] There is every indication that this synthesis was realised, but not without Pinchot's dedicated approach to continued learning and making meaning of his world. With typical dedication to resolving a conflict of apparent opposites, Gifford later found and read in a single sitting a copy of Henry Drummond's *Natural Law in the Spiritual World*. He was impressed and perfectly persuaded by Drummond's argument, declaring in his diary entry of 2 June 1891 that he had "needed just such a book".[102]

James Pinchot's suggestion to his newly graduated son that he consider the pursuit of forestry as a profession can be seen as a demonstration of Emerson's call to respect the student. Gifford's fascination with the world of nature had remained constant throughout his childhood and been encouraged and enabled by his parents. Nor had they neglected to ensure its balance by social, artistic and philosophical considerations.

---

98 ibid, Box 30, 1885.

99 Miller, *Gifford Pinchot*, 58.

100 ibid, 61.

101 Miller, *Gifford Pinchot*, 58.

102 Steen, *Diaries*,12.

Conclusion: Goyder and Pinchot

Gifford Pinchot's early education had led to his training as the first professional American forester, but it also led to an understanding by the young forester that a scientific knowledge of would be insufficient to secure public forestry in the United States (although he also later regretted abbreviating his European training). His learned skill was in bringing together the component parts of forestry and communicating this inside a larger narrative of a sustainable and equitable society. By the time of Theodore Roosevelt's presidency in 1901, Pinchot had played a leading role in laying the ground for the establishment in 1905 of the first Forest Service. His tenure as the first Chief enabled the "sustained growth of its personnel, an increase in its budget, and a heightened public awareness of its work and the importance of its mission".[103] His leadership depended as much on organisational and political insight as it did expertise in forestry science. Though a specialist in forestry, Pinchot later described the importance of forest conservation as part of a holistic conservation ethic essential to a civil society:

> Equality of opportunity, a square deal for every man, the protection of the citizen against the great concentrations of capital, the intelligent use of laws and institutions for the public good, and the conservation of our natural resources, not for the trusts, but for the people; these are real issues and real problems. Upon such things as these the perpetuity of this country as a nation of homes really depends. We are coming to see that the simple things are the things to work for. More than that, we are coming to see that the plain American citizen is the man to work for.[104]

Pinchot was a systems thinker (before the term was coined) with science training rather than a forestry science specialist. His ability to lead strategically across the boundaries of specialist knowledge, organisational structures and the complex mix of public needs can be seen as an application of the first principles of learning that had formed his early education. At the heart of his professional expertise was a well-honed system of values, learned early through an education of the head, heart and hand, though not named as such. These first principles of his education bore striking resemblance to his Australian counterpart, George Goyder. While not as transparent in Pinchot's learning, Swedenborg's and Pestalozzi's principles had had a powerful impact on those who influenced

---

[103] Miller, *Gifford Pinchot,* 156.
[104] Pinchot, *The Fight*, 36.

James, Mary and then Gifford. Spencer, Emerson and Thoreau had themselves learned from each man's profoundly humanistic philosophies.

In terms of the systems thinking Theory U, Pinchot's educational history offers extended insights into the formation of the "who" of his successful leadership of the innovation of formal forestry in the United States. Pinchot's educational history demonstrates the fundamental similarities between his and Goyder's learning to think across boundaries, of disciplines and organisations and national environments. The range of differences that characterised their social, political, economic and environmental circumstances, when unmasked, reveals the deep learning to which both men were exposed framed by their inheritance of the reflective reforming intentions of their parents. Each learnt similar capacities to construct a holistic vision of forestry as part of a sustainable society and to apply it adaptively within their different worlds. Each man's educational history reveals that the holistic vision of the component parts of their external world was informed by a capacity to construct knowledge across the internal human boundaries of what might be called head, heart and hand. In Pinchot's case, this was less direct and less prescriptive than in Goyder's. Pestalozzian and Swedenborgian philosophies, when present, were embedded in the thought and writings of reformers who influenced his parents' framing of his education, but the essential principles, crystallised in Pestalozzi's phrase, applied to the young Pinchot's learning and the parental influence in which this occurred.

Finally, this chapter has provided comparative evidence of another successful environmental leader and the connection between his early education and later adult success. In doing so, it has aimed to manage the tension in both historical and leadership research between the unique and the general. The comparison of Goyder and Pinchot has enabled a more persuasive case to be made for generalising the educational principles of leadership development. Part of the case rests in the differences in circumstances between the two leaders. Carr notes that by studying the evolution of events, or people, strikingly similar but occurring in "different historical milieu", and then comparing them, it is "easy to find the key to the understanding of this phenomenon".[105]

The concluding chapter to follow draws together the conceptual and empirical arguments of the book and provides an example of a direct application of the overall findings to a recent environmental "wicked"

---

[105] Carr, *What is History?*, 65.

problem as a final demonstration of a piece of "pragmatic" environmental history.[106]

[106] Dovers, "Sustainability", 21.

CHAPTER 11

CONCLUSION

The book set out to explore Dovers' proposition that "pragmatic" environmental history can be achieved by engaging with other "knowledge systems" while also constructing an historical narrative of "inherent interest" to the discipline. The other knowledge system has been management and organisation research, and specifically the sub discipline of systems thinking. A recent systems thinking theory, the Theory of the U, has framed the selection and analysis of the leadership of a wicked environmental problem of the past: the leadership of the innovation of scientific forestry in Australia and the United States.

Conceptually, the book has validated the Theory of the U by providing empirical evidence that innovative environmental leadership depends upon the leader's synthesis of external information filtered through an internal worldview of the public good. It confirms the Theory's proposition that the "who" is a key feature of leadership, rather than the much more commonly researched "what" and "how" of leadership. The book has also demonstrated the value to systems thinking of including rigorous historical research to expand the evidence base for testing the theory and uncovering potential solutions to leadership problems in doing so. In addition, the historical evidence has challenged the view of the past by most systems thinkers as either the location of the principle cause of present problems, or its complete irrelevance in the present for the future.

The empirical evidence of the historical case study has demonstrated that the complex nature (rather than specific content) of what are now called "wicked" environmental problems existed in the past. As such, their reconstructed narratives can offer models or case studies of their longitudinal process, outcome and leadership. These can offer insights to current problem solving. The focus of the case study was the impact of different leaders and the origins of their leadership in each man's personal learning history.

The contribution of this research to knowledge is the preliminary finding that systems thinking extends beyond the capacity to reflect across the boundaries of external and internal knowledge, as described in the

Theory of the U. Goyder's and Pinchot's stories suggest that externally, successful environmental leaders learn how to consider multiple external information sources. They have the capacity to synthesise political, social, physical and economic environmental information. Internally, they engage their "head, heart and hand" to evaluate that information against first principles of what a sustainable society looks like and needs. The complexity and uncertainty of a changing information base is continuously filtered through the certainty of these first principles of making meaning of their various environments. Diagrammatically, these processes may be represented as system loops at the top and the bottom and between the left side of the U image.

Reviewing the theoretical argument

Complex or "wicked" environmental problems are acknowledged as demanding a systems thinking approach for their formulation and solution. They are seen as unprecedented in kind, which necessarily limits their consideration to abstract thinking in the present for the future. This book has argued that the view of their unprecedented nature is a result of what management and organisation scholars have called the "presentism" of much of their field's research. The specific content of problems is new, but their complex and uncertain nature has parallels in the past. The assumption of unprecedented has denied the contribution which history can make to environmental wicked problem research.

Historians have long argued the practical value to present problem solving in their field. They have also suggested that a primary reason for its being overlooked in thought leadership has been the cultural dominance of the scientific over the humanist method of knowledge construction. They point to historical case studies as containing stimuli and lessons for current problem solving. Environmental historians also argue the practical possibilities for historical narrative, particularly in the engagement of the general public for changing problematic behaviour. But some environmental historians have called for a more direct pragmatic environmental history. They argue that the positive contribution history can make will be most effective when history directly engages with other knowledge systems. This is true, not least because such engagement captures another audience of scholars and in turn their practitioner audiences. In view of the ubiquity of management and organisational theory in the operation of contemporary institutions, this book combined management theory with environmental history to test this proposition. This pragmatic potential is compounded by the view of some management scholars that historical scholarship can add valuable context and complexity to their research.

Such scholars have begun to question the traditional "universalism and presentism" of their field. They ascribe these attributes to the dominance of the scientific method in their research culture, and believe that complex investigation and explanation has been often lost. In their meta-analysis of recent research using history, Usdiken and Kipping identified three categories of such use: "history *in* theory", "history *to* theory" and "historical cognizance". But systems thinking is not a field of management scholarship that has embraced history in the ways they describe. Where history has been employed it has been used to explain the origins of present problems. Some systems thinkers dismiss using history at all, arguing that the past at individual and collective levels serves as an impediment to thought and action. Their temporal focus is on leaders' ability to lead "from the emerging future" to achieve truly innovative change. In view of the central role that systems thinking is agreed to play in environmental wicked problems, this book selected a theory posited by leading systems thinkers. The hypothesis was that history has a constructive contribution to make to addressing such problems. The Theory of the U was devised and refined by Peter Senge and colleague Otto Scharmer. Senge has used history to explain present problems. Scharmer has argued that history is an obstacle to innovative leadership. Both agree that the "blindspot" of leadership research has been the "who" of leadership, with the "what" and the "how" of leading much more dominant in the research. The Theory proposes that a leader's engagement with external sources of knowledge must be filtered through an internal knowledge base in order to produce sustainable and effective results.

The book used the Theory to frame and select an environmental wicked problem of the past, the innovation of scientific forestry in Australia. It has demonstrated that there is a place for "history *to* theory" that expands the evidence base for the validation of the Theory. In addition, it has demonstrated that personal learning histories can deepen the understanding of how the "who" of leaders is developed. The engagement of the "knowledge systems" of environmental history and systems thinking has produced a piece of "pragmatic" environmental history.

It has also produced a historical narrative, what Dovers has called history of "inherent interest" for the general public. This too has practical application in influencing changed behaviour, as Cronon argues. The narrative has a self-contained value through its accessibility not only to citizens but also to specialists, as Griffiths argues. Reading the story of complex environmental problems and comparing the effective approaches of Goyder and Pinchot with those of Finniss and Brown suggests possibilities for new ways of thinking and acting in terms of the environment.

The story of the introduction of scientific forestry and its leadership offers a longitudinal case study, unavailable for present data, as stimulus for the consideration of complex environmental problems in the present for the future. The generalisability of the conclusion that early learning principles can be linked to adult innovative capacity is made more robust by the extension of the research to include the educational history of the leadership of American forestry.

Reviewing the empirical argument

The findings from the book support the view that environmental wicked problems have precedents. The descriptors commonly used in the management literature to describe such problems are readily applicable to the historical case study. Such descriptors are summarised in the Australian policy document "Tackling Wicked Problems". The parliamentary and public debates that attended the protracted introduction of scientific forestry in Australia demonstrated that the causal problem was "difficult to define", that it was "socially complex", and that attempts at solution could "lead to unforeseen consequences". The implementation of the service further demonstrated the "interdependencies", the lack of "clear solution", and significantly the "need to change behaviour of citizens and of government for long-term sustainability", rather than "relying solely on increasing the regulatory environment". An examination of the thought and action of key leaders in the implementation of the innovation through the Forest Board highlighted their different approaches. Unlike Goyder, Finniss and Brown were unable, or unwilling, to recognize the complexities of the problem at the scientific, social, political or economic levels. An examination of the educational histories of each man suggested reasons why Goyder's produced more successful results. He had been taught to make meaning of the world in ways that were not content-laden. Instead, they provided a framework that enabled continuous learning for action in uncertain and evolving circumstances. His education was powerfully supported by his parents' reasoning and values, encouraging not only the apprehension of external phenomena across the social, physical, economic and political boundaries but also against an evolving self-knowledge and worldview. He learned to know himself and his environments across the internal boundaries of "head, heart and hand".

The comparative educational histories of Finniss, Brown and Goyder provide exploratory findings for arguing not only the importance of the "who" of leadership, but also how it is developed in the adult leader. The Australian case study is rendered more robust when compared with another successful leader in a different time and context. Gifford Pinchot, as the much more celebrated "father" of the United States' forestry, offered the possibility of arguing more strongly the generalisability of the findings from Goyder's case study. The evidence highlighted the significant disparity in the social, political, economic and environmental circumstances of Goyder and Pinchot. Finally, it demonstrated significant deeper similarities in the educational principles that shaped their childhood learning. These first principles suggest a framework for the development of the "who" of effective leaders for environmental wicked problems. They could be deployed at any stage of leadership education but are, as the Victorian Bushfire Royal Commissioners noted, most effective when used in the education of the young for environmental wicked problem seeing and solving.

The following section summarises the findings of the book before applying these to the two recommendations on education and research contained in the Victorian Bushfire Royal Commission report, a recent example of a wicked environmental problem. This is offered in final and direct demonstration of the value of these "pragmatic" environmental historical case studies.

Political leadership of an environmental wicked problem

By 1870, Goyder and other prominent South Australian leaders, notably Friedrich Krichauff and Richard Schomburgk, were decided on the essential importance of public forestry for progressing and sustaining settlement in the new colony. But it took another three years to legislate to encourage private forestry and a further two for an Act formalising the first government forestry authority in Australia. However modest the final result when measured against the desires of its leading proponents, the innovation of scientific forestry was formalised in South Australia, and Australia, in 1875. What were the attributes of leadership that enabled this success? And what were those that impeded it?

Goyder's ability to envisage the big picture of the sustainable settlement of an immigrating human population in the physical environment of the colony was vital to the legislative success. Taken together, his 1870 reports on the Victorian Land Laws Acts and on forest reserves demonstrate his established view of the necessity of forestry as part of the total tapestry of sustainable settlement. The former is testimony to his framing of the human and natural landscape by an established philosophy of what a civic society looked like. His practical efficiency was informed by a holistic value system, and his notion of a civic society in which the distribution of land occurred after survey and after the conception of what a sustainable community looked like were on a continuum with the founding philosophy of planned South Australian settlement.

Goyder was not empowered in this evolving democracy to make absolute decisions for land allocation for forestry. In the houses of parliament, the large picture of civic settlement and forestry's place within it were subject to the democratic process. Here the complex, multi-faceted mess of wicked problems was clearly on display. Collaboration with Friedrich Krichauff and Richard Schomburgk was vital to achieving the final legislative outcomes. Even though these fellow champions of forestry were at odds with Goyder on elements of the science of forestry they held on to the shared necessity of scientific forestry throughout the five-year-long debates. Their perseverance kept the matter of forestry before parliament, prodding members' general but inert approval of forestry into an evolving acceptance with a final legislative mandate to support it. In this early stage of orchestrating forestry's acceptance in government circles, it mattered less that Krichauff and Schomburgk's promises for climate change were accurate than that they supported the innovation, nor did it matter that Goyder's predictions of the revenue generating power of plantation forestry for the colony were unrealistically high. The pressing requirement of thought leadership in the first half of the decade was to sustain the large argument that deforestation needed regulation and that afforestation was broadly beneficial. Parliamentarians' engagement with the detail of debates rarely challenged these foundational premises and, though tedious in their protraction, contributed to raising the profile of forestry in the colony over time through their reporting in the public press.

Where there was excessively detailed debate was in the Forest Board bill. If Goyder and Krichauff judged the timing right for the introduction of the bill to secure forestry, Goyder's misjudgement lay in the delivery of a highly detailed series of clauses defining the management operation. Rather than eliminate the need for members of parliament to concern themselves with the operational minutiae of forestry oversight, the articulation of the fine workings gave rise to a multiplicity of views, each wanting to be heard. Goyder misread parliament's willingness, in 1874, to forefront the big picture. Parliamentarians' uncertainty of the necessity of forestry had been settled but was overtaken by an uncertainty of how to implement the innovation. Endless circularity could have resulted but was brought to a close by the Commissioner, probably on Goyder's advice, suspending Standing Orders after an eighteen month-long debate.

It had taken the firm grasp of the big political, social, economic and environmental picture in Goyder's 1870 reports, supported by a shared vision of forestry by other credible proponents with a voice, for the first formal step towards sustainable forestry in Australia to be taken. Significantly more important in this achievement of persuasion had been an understanding of the political rather than natural landscape. When this faltered, it took accepting both the need for protracted debate and for exercising the available tactical intervention to bring this to a satisfactory end.

The Forest Board's leadership of an environmental wicked problem

Goyder's leadership of the second stage of implementation was more direct and more visible than it had been in the parliamentary debates. A reasonable expectation may have been that a combination of a small, specialist group and common desire to implement forestry would produce smooth and productive outcomes. Instead, the conflicts intensified. The antagonisms within the Board were galvanised by the appointment of Brown as Conservator. If the principal impediment to enacting legislation had been sustaining an interest in action among parliamentarians, the inverse was true of the implementation phase. The impatience of Board members to see forestry realised on a grand and exclusive scale led to the implosion of the Board just over three years after Brown's arrival. Yet the life of the Board, the public profile its work on behalf of forestry had achieved, the promotional zeal of Brown, and even the publicity which the fierce conflicts attracted in the newspapers all contributed to the acceptance by 1882 of public forestry as the legitimate business of government. The controversial seven-year life of the Board and its eventual demise were parlayed by Goyder to secure the establishment of the first Woods and Forests Department in Australia, under the directorship of Brown, but within the portfolio of Goyder's Lands Department.

Goyder's holism, or what might now be termed systems thinking, was vital to reaching this secure position for forestry twelve years after its introduction to parliament in the colony. A key element of that success was recognising the importance of opening up the evolution of the innovation to the education, scrutiny and engagement of the whole community through Goyder's invitation to the press to attend and report on Board meetings. During the publicised conflict, Goyder's views remain disinterested and understated. For him, the ridicule in the press and hostility within the Board was not a matter of ego. In the case of the scientific theory that linked forests to the cause of rain, he knew from his iterative application of the scientific method in the field that it was wrong science. In the case of the multiple use of forest reserves, he knew from an economic, political and social perspective that the exclusion of all activity in the forests other than forestry privileged forests in unbalanced ways, with potential risks to physical and economic safety.

Brown and Finniss learned to see their worlds in isolated parts of the whole, and to focus their vision and action accordingly. Goyder, by contrast, learned to see his world as continuously adapting parts in a dynamic relationship with the whole. The whole needed to be understood clearly in order for the part and parts to make sense, and for their relationship to the whole to be grasped and acted upon. The "who" of each man's learned leadership was differently shaped and, as Senge and Deming note, had its foundation in childhood.

The "who" of leadership

Deming's and Senge's shared view is that the often poor management of organisations, public and private, is mirrored in the educational institutions that help to develop its leaders and workers. The arbitrary command-and-control nature of this educational system had had a generally deleterious effect on the humanity and effectiveness of both. This view suggested the value of examining the educational histories of the key leaders of forestry's implementation as one way of examining Senge and Scharmer's declared "blindspot", the "who" of the leader.

Foundational to the "who" of Finniss' leadership was his inability to operate in the absence of external authority. He had been deprived of a formal and informal learning environment where he was able to develop deeply reflective and critical modes of thinking, feeling, and experiencing his world. What had become embedded instead was a hierarchical style of leadership, clear in his view of organisational authority and in the authority of knowledge. He saw leadership as a contest of power. Because Finniss lacked a developed capacity to continuously learn and adapt to a new and changing environment, and had not been exposed to a process that developed such a capacity, his criteria for independent judgement, and certainly for creative innovation, was severely constrained. Instead, he exercised his personal authority within the Board in destructive ways. He forged the adversarial and finally terminal dynamic between Board members and Goyder, attaching the absolute authority of the Board to decide and act to the certainties of science which Brown embodied.

If Finniss' pre-professional life trajectory can be seen as framed by his father's military career, Brown's too was closely entwined with his father's. Where Finniss's formal schooling and the associated informal culture of English boys' public school life mirrored the military life, John's formal schooling was unremarkable when viewed against the educational influence of his father. At local schools, he received a foundation in subjects which his father named as basic grounding for the development of forestry knowledge. By age fifteen, when he left school to become his father's apprentice, the exclusive source of his learning became his father. Brown followed his father in learning the craft of writing about forestry and joining the associations in which his father played a leading role. He also accompanied his father as he sought further elevation of the status of forestry through an international network of similar research-based associations. The father and son were particularly involved with early North American forestry. By the time of his appointment to the position of Conservator, Brown had published papers on North American forests and was to become a member of the prestigious Linnaean Society.

Unlike Finniss, Brown's learning environment to the age of adulthood was marked by a supportive familial context. His seemingly inevitable trajectory toward professional forestry was surrounded by a heady mix of practical and intellectual excitement about the emerging field. The prodigious work ethic Brown displayed in South Australia in the late 1870s demonstrated the son's absorption of the father's passionate evangelism on behalf of forestry. Brown's zeal lacked the destructive and negative impulses of Finniss. It contained much of his absolute belief in his, and therefore the Board's, authority to direct the implementation of forestry. In Brown, there was a youthful version of the elder statesman Finniss' paternalism and hubris, a narrowness of vision for antipodean forestry that transposed its theory and practice from the matured field in Europe. As his father's loyal protégé Brown was even more persuaded of the universal application of James' forestry science whatever the particularities of the physical or social environments. Brown's certainties were entwined with a filial loyalty to his father. He had not been taught how to learn forestry contextualised by the synthesis of physical, political, social and economic environments in which it was to be practised. His absolute conviction of the rightness of the scientific theory of forests causing increased rainfall, and the exclusion of all human activity apart from forestry from forest reserves can be seen as much a belief in his father's infallible singular vision as it was in the absolute truths of scientific theory. Nevertheless, the constructive, supportive context of his learning enabled him to adapt to the next stage of forestry's innovation in the colony and become the head of the new department of woods and forests, and to go on to lead the innovation of forestry in other Australian colonies. His uncritical zealotry matured while Finniss's led to destructive and self-destructive consequences.

At the centre of the "who" of Goyder's leadership was his capacity to think and act in complex wholes rather than specialist parts. But this capacity did not deny rigour. This was evident in his view of the theory of forests and rainfall. He knew from direct and repeated observation of the physical environment of the colony both the necessity and limits of forestry's promise. He knew the contingent truth of scientific theory immersed as he was in the scientific method of discerning environmental patterns. Equally, he knew the contingent nature of human behaviour and the uncertainties that attended the acceptance of much of his work. The certainties he held were of clear principles for settling a civic society and a learning process that expected uncertainty and complexity.

Goyder's educational history is no less a story of a son's trajectory framed by his father's professional endeavours, but in more expansive ways than were experienced by Finniss or Brown. Both his mother Sarah and father David figured prominently in the young Goyder's learning, and as Sarah shared some of David's professional life so too did he share some attributes stereotypically assigned as feminine. His deep compassion and nurturing sensibilities are on display in his autobiography and in his work with young children, which had the same large social reform purpose that had been the impetus for Pestalozzi's educational method. The partnership of Goyder's parents was shaped by the boundary crossing principles of both Pestalozzi and Swedenborg. They thought and acted across social, political, economic and physical circumstances as well as across external and internal boundaries. Their social agendas were informed by a deep humanism. In an age where the Enlightenment was leading to the production of specialist and materialist disciplines, they retained a synthetic and integrated worldview. It is not surprising that the "practical realism" of these worldviews, which allowed for subjectivity as well as objectivity in matters dealing with human behaviour, produced in David and Sarah Goyder a humanism that transcended the usual gender divisions of domestic and public endeavours. While the same shared professionalism of husband and wife was not evident in the structure of George Goyder's own life, this is hardly surprising given the exigencies of living in a new settlement and Goyder's exploratory and formative role in that. But the same compassionate humanism which underpinned his parents' domestic arrangements is evident in his ideas and actions to ensure the creation of a sustainable civil society in the colony.

Goyder had been taught to learn using head, heart and hand in an integrated way, a way that was inclusive of all of the dimensions of human knowledge construction. He was taught to do so in order to enhance and improve society in as equitable fashion as possible, maximising the opportunities for an individual's dignity and self-improvement. Respectful treatment of natural resources, including the land and its use, were fundamental to the achievement of social ideals. Goyder's insistence on the view that land must continue to be surveyed before settlement was essential to the allocation of land at the centre of regional communities for public services, such as schools, so that all settlers would have access to the civilising and humanising resources needed in an evolving society. For him, this was both a pragmatic and heartfelt use of the natural resources for social, political and economic ends.

For Goyder, forestry was also a pragmatic requirement of sustaining and developing a new society, but its claims to the land had to be balanced by the competing needs of that emerging society. It should not and could not be considered in isolation. Goyder had been taught to read parts inside their wholes and in context. This was not a formula for infallible leadership, but for continual adaptation towards a goal determined by an assessment of the public good. His view of what constituted the public good was determined by considerations that straddled head, heart and hand. Extending the scope of the book to include an educational historical narrative of Gifford Pinchot argued a stronger case for generalising the findings of Goyder's and Pinchot's successful environmental leadership.

Comparing the findings to another successful environmental leader

Pinchot's eventual success in establishing the first US Forest Service in 1905 was the culmination of efforts began in the 1870s, when forestry was first mooted as a necessity for the sustainable management of the nation's natural resources. Several scientifically trained, mainly German, foresters had preceded Pinchot in attempting to professionalise forestry and elevate its status as acceptable public business. Pinchot's scientific training was arguably less comprehensive than his predecessors but his ability to read forestry holistically within a total picture of society and to contextualise it within a whole conservation ethic that framed an aspirational society was new. The resonance of Pinchot's approach with Goyder's is powerful.

But the similarities of their early learning require a view below the surface of striking difference that mirrored the different landscapes they oversaw. Their generational difference, opposite physiques and eloquence were marked as were the wealth and social position of their respective families. Nevertheless, at the level of first principle, they shared much, and this began with their parents. James and Mary Pinchot were young adults when the nation was overwhelmed by the catastrophe of civil war. Their reflective reformist views were informed by artistic and emotional sensibilities that responded to the national need for post-war reconstruction, including the shaping of the education of Gifford (Gifford's naming after Sanford Gifford, a civil war artist and friend of the family, stood in contrast to the other children each named after another family member). Although Mary did not share a professional partnership with James, her capacity for independence was evident in more than her separate, inherited wealth. She spent protracted periods in her children's early years living away from James and her extended family in Paris, nurturing her own fragile health possibly caused by the death of their first daughter Lucy. By the mid-1870s, when Gifford was ten years old, James had retired from professional employment and they shared the responsibility of raising of their three young children.

Gifford's early and persistent love of the world of nature was actively encouraged, informed by James' reflective considerations of educational philosophy and supported by contact with leading thinkers of the day such as Spencer and the writings of Thoreau and Emerson. The child-centred, experiential method of learning that was a feature of Goyder's early education also framed Pinchot's. Though the direct influence of Pestalozzi and Swedenborg is not as apparent, each of these men figured in Gifford's learning through their influence on Spencer and Emerson.

The approach to learning their role in the world that had characterised Finniss's and, in more subtle ways, Brown's education, and depended upon the child being moulded to fit a predestined parental outcome, was missing from both Goyder's and Pinchot's methods of instruction. Pinchot had learned from his parents and through his own maturing wisdom to integrate what were seen as quarantined fields of enquiry. So, he blended the aesthetics and the metaphor of forestry, taught by Yale's Professor of Fine Arts, John Weir, and painted by his godfather, Sanford Gifford, with the science and politics of forestry. He blended the spirituality of forestry with the economics of forestry. He saw forestry itself within the total picture of a progressive and aspirational society, which was his inheritance and to which it was his duty to contribute in the ongoing refinement of the republic.

In the end, the differences of country, generation, wealth, position, and society that clearly separated Goyder and Pinchot, and in which they operated as both children and adults, had a less profound effect on the "who" of their leadership than the similarities of affective and intelligent parental engagement and the educational principles that framed and continued to shape their learning. Each child learned to consider their natural, social and physical environments as a whole first and as separate parts in the service of the whole second. Each child learned to think not only with their head but also with their heart and with their senses, so that later the "what" and "how" of forestry were determined by a holistic approach to its practice and to its role in civic society, filtered through a holistic and continuously maturing knowledge of "who" each man was.

Applying the historical findings to the Theory of the U

The book has presented a case study to contribute to the call for "pragmatic" environmental history achieved through the engagement of the discipline with other knowledge systems. It has brought together literature and theory from management with the historiographical literature and a narrative case study of the thought and action leadership of the "wicked" environmental problem of innovating forestry. The choice of the former is based on the premise that management literature occupies a dominant place in policy and practice in the public as well as private sector and systems thinking, from which the Theory of the U derives, plays a prominent role in addressing current environmental wicked problems. The literature of environmental wicked problems and the Theory of the U were used to select the historical topic and evidence and to frame the research questions to be asked of the archival sources. The pragmatic outcomes of the book were to test the validity of the Theory using historical evidence and to suggest the elements of the "who" of successful leadership and how they might be learned. The book supports the view that by combining the methods of environmental history with those of other disciplines, a dialogue is established that bridges the different cultural languages to make history more intelligible to and applicable by another field.

The book framed its findings in terms defined by management scholars Usdiken and Kipping, who describe how some in their field are using "history *in* theory" in which "history serves as evidence to develop, modify or test theories". The findings of the educational histories of the four leaders of the innovation of forestry in Australia and in the United States support the theory that the "who" of leadership is core to the successful steering of complex environmental problems toward stable resolution. They also suggest that of central importance to successful decision-making is a deeply considered and holistic vision of the aspirational society to which the innovation will contribute. The learning histories of environmental leaders of the past offer one way of researching the replicable principles for educating or re-educating leaders in the present for the future. Such histories may be of recent or more distant leaders of the past. The particular value of those more distant in time is that they may be clearly seen with the detached view of hindsight and inside a context that is at once familiar but lacks the potentially obscuring lens of immersion in the murkier present. Given the accepted view by researchers that leadership is a process requiring longitudinal study, problems and their leaders from the past offer a powerful opportunity to examine both. Beyond the value to researchers of leadership or wicked problems, the historical narratives presented here also offer the kind of practicality that Cronon and Griffiths champion. They provide accessible stories for the general public interested in the past itself and interested in using the lessons of history as stimulus and analogy for thinking through environmental problems with which they are confronted in the present.

The application of management theory to historical evidence is one way of encouraging the greater use of historical thinking in the consideration of present problem solving and leadership. The use of historical evidence for the Theory of the U offers a way of expanding the evidence base for systems thinking, which has tended to invoke the past primarily to explain the source of contemporary problems. The book has made a strong case for the inverse: that rigorous historical scholarship which pays detailed attention to context and the "who" of historical actors can support solution finding for systems thinking research.

Applying findings to the Victorian Bushfire Royal Commission's recommendations

Specifically, the exploratory findings of this case study environmental history of leadership of complex problem-solving can be applied to the two Recommendations of the Victorian Bushfire Royal Commission, the concrete example of an environmental wicked problem of the present which helped introduce the book. In doing so, the book concludes with a demonstration of direct applicability of its findings.

The VBRC proposed two recommendations that concern new knowledge and new education: Recommendations 6 and 65. The first reads:

> Victoria lead an initiative of the Ministerial Council for Education, Early Childhood Development and Youth Affairs to ensure that the national curriculum incorporates the history of bushfire in Australia and that existing curriculum areas such as geography, science and environmental studies include elements of bushfire education.

The second reads:

> The Commonwealth establish a national centre for bushfire research in collaboration with other Australian jurisdictions to support pure, applied and long-term research in the physical, biological and social sciences relevant to bushfires and to promote continuing research and scholarship in related disciplines.

While history was named as a key discipline in the first recommendation, that focus was dropped in its final implementation. The absence of the humanities completely from the second recommendation automatically includes history. These recommendations and their implementation reinforce the view by the management scholars and applied historians with which the book began: that positivist or scientific methods and the disciplines which use them have captured public management and policy making because these methods claim superior objectivity and quantifiable large scale findings with the promise of dramatic and unprecedented change for unprecedented problem-solving. But the complexities of human behaviour defy such clinical, decontextualized examination. The richness of human behaviour requires the same complexity of investigation to interpret in support of understanding in the present for future action. History with its humanist methods can contribute to that by presenting whole pictures of human systems. These can sit alongside disciplines that privilege "universalist and presentist" methods to provide an expanded research and educational base from which to develop future environmental leaders. The book demonstrates that history should be included in the education and research reform agendas for very practical public outcomes, as the Royal Commissioners first intended.

Applying the findings to the leadership of environmental "wicked" problems

More generally, the findings of the book can be applied to the leadership of one of the wicked problems named by the Australian Public Service Commission: Climate Change. The political portrayal of this pressing and complex issue often takes the form of a binary divide between climate believers and climate deniers. Opposing scientific evidence is regularly enlisted in substantiation of each position. This present scenario resonates strongly with the debates within the Forest Board in nineteenth-century South Australia. Brown, Finniss, preceded by Krichauff and Schomburgk, were convinced by the scientific theory of the relationship between forests and rainfall that plantation forestry would transform the climate of South Australia to enable sustainable agriculture much further north than Goyder deemed safe. Brown embodied this scientific theory, the social and economic attraction of which was compelling in a colony hungry for expansive, sustainable settlement. Had his reductionist and singular view prevailed large scale planting is likely to have occurred in the north, and the mixed use of the developing plantation forests of Bundaleer and Wirrabara eliminated in favour of exclusive plantation forests. The inevitable failure of the first is now self-evident, as it was to Goyder at the time. The potential damage to the growing economy, the social fabric and even the sustainability of the forests themselves is also now clear. The mixture of skepticism, complex thinking, compromise and transparency of motive demonstrated by Goyder instead produced a more balanced outcome. In structural terms, a Department of Woods and Forests which has endured for over one hundred and forty years. Had Brown's singular ideology of forests and climate prevailed the economy and viability of human settlement would have been put at substantial risk. The public persuasion of the importance of forestry to the general public may have been delayed by the time it took to rectify the damage to trust and landscape of Brown's misguided implementation.

Goyder and Pinchot led the establishment of enduring forestry services in each of their nations. Their learning to learn to think and act in non-reductionist ways in the midst of the inevitable complexities of environmental problem-solving offer models of learning principle not only to political and policy leaders of today's "wicked" problems, but to the general public whose democratic right and responsibility is to continue to challenge any oversimplified interpretation of their multi-faceted environment.

BIBLIOGRAPHY

Primary historical sources

Parliamentary papers:
South Australian Parliamentary Debates (SAPD).
South Australian Parliamentary Proceedings (SAPP).

Newspapers:
Adelaide Observer
The Advertiser
The Argus
The Jamestown Review
The South Australian Register

State Records of South Australia:
GRG 15772, Minutes of memoranda relating to matters raised by the Board - Secretary, Forest Board, later the Conservator of Forests, 1876-83
GRG 16/1, Rough Minutes – Forest Board, 1875-81
GRG 16/2, Subjects for consideration at the meetings of Forest Board, 1878-80
GRG 16/3, Letters sent - Conservator of Forests, 1878-80
GRG 16/7, Unregistered correspondence, Forest Board Office, 1877
GRG 16/8, *Woods and Forests Newspaper Cuttings*, State Records of South Australia.
GRG 16/11, B.T. Finniss papers - Forest Board 1875 -1882, State Records of South Australia.
GRG 16, Woods and Forest Department, 1883-1997
GA246, Woods and Forest Department, 1883-1997
GRS/ 116168, Pastoral run surveys, 1860-90, Surveyor General's Office
GRG 35/1, Correspondence files (`CL I` or `CLO` dockets) - Crown Lands and Immigration Office, 1856-1917
GRG 35/1/00000/20, File 194/1865 G.W. Goyder, Surveyor General, 1865-1965
GRG 35/2, Correspondence files (`SGO` files) - Surveyor General`s Office, 1838-1917

GRG 35/19, Private letter book - Surveyor General`s Office, 1868-1918
GRG 35/234, Completions of purchase of land under credit agreement - Surveyor General`s Office, 1872-1913
GRG 35/249, Circulars and memoranda (`general orders`) issued - Land Office; Surveyor General`s Office; Lands and Survey Department, 1839-1940
GRG 35/286, Field note books of G.W. Goyder as Assistant Surveyor General, later Surveyor General, 1856-93
GRG 35/329, Atlas of the principal portions of the settled districts in the province of South Australia - Surveyor General`s Office, 1877-8
GRG 35/362, Abstract of land grants prepared, Surveyor General's Office, 1871-82
GRG 35/376, Area book, Surveyor General's Office, 1861-70
GRG 35/522, Alphabetical index to pastoral leases, Surveyor General's Office, 1851-1892
GRG 35/580, Crown Lands rent books - Surveyor General`s Office, 1868-83
GRG 35/587, County maps - Surveyor General`s Office; Lands and Survey Department; Lands Department, 1845-1958
GRG 35/655, Diary kept by Surveyor General G.W. Goyder - Northern Territory Survey Expedition 1868-1870

Diaries and letters:

Gifford Pinchot Papers, Library of Congress, Washington DC, Boxes 39 - 76f.

Legislation:

An Act to Encourage the Planting of Forest Trees, no. 26. 1873.
Forest Board Act no. 8, 1876.
Crown Laws Consolidation Act no. 86, 1877.

Other:

Missouri State Archives Finding Aid 3.23 Office of Governor John Smith Phelps, 1877-1881.

Government reports

Australian Public Service Commission, *Tackling Wicked Problems: A Public Policy Perspective,* 2007, Canberra; Australian Public Service Commission. *Changing Behaviour: a Public Policy Perspective.* http://www.apsc.gov.au/publications07/wickedproblems.pdf.

The 2009 Victorian Bushfires Royal Commission final report. http://royalcommission.vic.gov.au/Commission-Reports/Final-Report.html.

OECD Centre for Educational Research and Innovation (CERI), 'Educating for Innovative Societies' Conference Summary Report, 26 April 2012, Paris, France.

Theses

Carrs, Nancy. 1981. "Grey Towers 1884-1890: A Social, Architectural, and Decorative Arts History of James Wallace Pinchot's Country House in Milford, Pennsylvania." MSc. diss., Columbia University.

Hirst, John. 1963 "G.W. Cotton and Workingmen's Blocks". B.A. (Hons) diss., University of Adelaide.

Husmann, J. 2004 Transplantations: a Comparative History of Afforestation in Nebraska and South Australia 1870s- 1940s. PhD diss., The University of Nebraska, Lincoln.

Payne, Pauline., "Dr. Richard Schomburgk and Adelaide Botanic Garden, 1865-1891". PhD diss., The University of Adelaide.

Sheldrick, Janis. 1999 "Nature's Line: George Goyder Over South Australia". PhD diss., Deakin University.

Presentations

Horn, Robert. 2001 "Knowledge Mapping for Complex Social Messes", A presentation to the "Foundations in the Knowledge Economy" at the David and Lucile Packard Foundation, July 16.

Rainsford, Tamara. 2009 "Back to the Future: Educating the Renaissance Scholar", unpublished presentation to the Institute for Phototonics, University of Adelaide, September 16.

Scharmer, Otto. 2011, "Leading from the Emerging Future", Paper presented to Federal Ministry for Economic Cooperation and Development, November 13, Berlin.

Books

Appleby, Joyce, Hunt, Lynn and Jacob, Margaret. 1994. *Telling the Truth about History*. New York: W.W. Norton & Company.

Balint, Peter, Stewart, Ronald, Desai, Anand, and Walters, Lawrence. 2011. *Wicked Environmental Problems: Managing Uncertainty and Conflict*. Washington DC: Island Press.

Booth, Jenny and Borrow, Keith. 2001. *B.T. Finniss: First Premier of South Australia*. Adelaide: published by authors.

Brown, John.1881. *A Practical Treatise on Tree Culture in South Australia*. Adelaide: Government Printer.

Brown, James.1851. *The Forester: Plain and Practical Directions for the Planting, Rearing, and General Management of Forest Trees*. Edinburgh: W. Blackwood & Sons.

Carr, Edward. 2008. *What is History?* Camberwell: Penguin Group.

Cronon, William. 1991. *Nature's Metropolis*. New York: W.W. Norton & Company.

Cronon, William. 1996. *Uncommon Ground: Rethinking the Human Place in Nature*. New York: W.W. Norton and Company.

Dargavel, John, and Johann, Elisabeth. 2013. *Science and Hope: a Forest History*. Cambridge: White Horse Press.

Dovers, Stephen. 2005. *Environment and Sustainability Policy*. Sydney: Federation Press.

Dovers, Stephen and Hussey, Karen. *Environment and Sustainability: a Policy Handbook. 2013*. Sydney: Federation Press.

Emerson, Ralph.1910. *Representative Men*. London: The Waverley Book Company.

Avery, Kevin, Kelly, Franklin, and Conway, Claire. (eds). 2003. *Hudson River School Visions: The Landscapes of Sanford R. Gifford*. New York: Metropolitan Museum of Art.

Gladish, Richard. 1968 *History of New Church Education Vol. 2, 1800-1968*. Bryn Athyn: The Academy of the New Church.

Goyder Kerr, Margaret. 1971. *The Surveyor: The Story of the Founding of Darwin*. Adelaide: Rigby.

Goyder, David. 1857. *My Battle for Life*. London: Simpkin, Marshall & Company.

Goyder, George. 1890. *Report by the Surveyor-General upon the Disposal of Public Lands of South Australia*. Adelaide: Government Printer.

Griffin, Patrick, McGaw, Barry, and Care, Esther (eds). 2012. *Assessment and Teaching of 21st Century Skills*. New York: Springer, Dordrecht.

Hirst, John. 1973. *Adelaide and the Country 1870–1917*. Melbourne: Melbourne University Press.

Home, Rod, Lucas, Arthur, Maroske, Sara, Sinkora, Doris, Voight, Johannes and Wells, Monika., (eds), *Regardfully Yours: Selected Correspondence of Ferdinand von Mueller, Volume III: 1876-1896*. Bern: Peter Lang.

Hughes, Thomas. 1857. *Tom Brown's School Days*. London: Macmillan.

Isaacson, David. 2007. *Einstein: His Life and Universe*. London: Simon and Schuster.

Jonsson, Inge. 1971. *Emanuel Swedenborg*. New York: Twayne Publishers, Inc.

Lewis, Norman. 1975. *A Hundred Years of State Forestry: South Australia: 1875-1975*. Adelaide: Woods and Forests Department.

McCullough, David. 2013. *The Greater Journey: Americans in Paris*. New York: Simon and Schuster.

McGeary, M.Nelson. 1961. *Gifford Pinchot: Forester-Politician*. Princeton: Princeton University Press.

Meinig, Donald. 1962. *On the Margins of the Good Earth: The South Australian Wheat Frontier 1869-1884*. London: John Murray.

Miller, Char. 2012. *Public Lands, Public Debates*. Corvallis: Oregon State University Press.

McClure, James and Looney, J. Jefferson (eds). 2009. *Jefferson Papers Online*. https://jeffersonpapers.princeton.edu/about-us/jefferson-papers-online. Accessed November 30, 2014.

Oosthoek, Jan. 2013. *Conquering the Highlands: a history of the afforestation of the Scottish uplands*. Canberra: Australian National University Press.

Pestalozzi, Johann. 1898. *How Gertrude Teaches Her Children: An Attempt to Help Mothers to Teach their Own Children*. New York: C.W. Bardeen.

Pike, Douglas. 1967. *Paradise of Dissent: South Australia 1829-1857*. Melbourne: Melbourne University Press.

Pinchot, Gifford. 1910. *The Fight for Conservation*. Seattle: University of Washington Press.

Powell, Joseph. 1988. *An Historical Geography of Modern Australia*. Cambridge: Cambridge University Press.

Roosevelt, Theodore. 1913. *An Autobiography*. New York: Macmillan.

Scharmer, Otto. 2009. *Theory U: Leading from the Future as it Emerges*. San Francisco: Berrett-Koehler.

Senge, Peter, Jaworski, Joseph, Scharmer, Otto, and Flowers, Betty 2005. *Presence: Exploring Profound Change in People, Organizations and Society*. London: Nicholas Brearley.

Senge, Peter, Cambron-McCabe, Nelda, Lucas, Timothy, Smith, Bryan, Dutton, Janis, and Kleiner, Art. 2000. *Schools that Learn: A Fifth Discipline Fieldbook for Educators, Parents, and Everyone Who Cares About Education*. London: Nicholas Brealey.

Sheldrick, Janis. 2013. *Nature's Line: George Goyder, Surveyor, Environmentalist, Visionary*. Adelaide: Wakefield Press.

Sherman, William. 2006. *The Memoirs of General W. T. Sherman*. https://www.gutenberg.org/files/4361/4361-h/4361-h.htm. Accessed June 13, 2014.

Sigstedt, Cyriel. 1971. *The Swedenborg Epic:the Life and Works of Emanuel Swedenborg*. New York: Bookman Associates.

Silber, Kate.1973. *Pestalozzi: The Man and His Work*. New York: Routledge and Kegan Paul.

Spencer, Herbert.1861. *Essays on Education & Kindred Subjects*. New York: Dutton.

Spencer, Herbert. 1893.*The Study of Sociology*. New York: D. Appleton.

Steen, Harold. 2001. *The Conservation Diaries of Gifford Pinchot*. Durham: Forest History Society.

Stretton, Hugh. 1969. *The Political Sciences*. London: Routledge and K. Paul.

Tosh, John. 2010. *The Pursuit of History*. London: Routledge.

Tyrrell, Ian. 1999. *True Gardens of the Gods: Californian-Australian Environmental Reform, 1860-1930*. Berkeley: University of California Press.

Wakefield, Edward. 1829. *A Letter from Sydney*. Letchworth: Temple Press.

Worster, Donald. 2008. *A Passion for Nature: The Life of John Muir*. New York: Oxford University Press.

Wulf, Andrea. 2012. *Founding Gardeners: The Revolutionary Generation, Nature, and the Shaping of the American Nation*. New York: Random House.

Articles & book chapters

Ackoff, Russell."The Art and Science of Mess Management." Interfaces. 11(1) (1981): 20-26.

Allen, Gerald and Gould, Ernest. "Complexity, wickedness and public forests." Journal of Forestry 84(4) (1986): 20-24.

Anderson, Robert. "Education and the State in Nineteenth Century Scotland." Economic History Review. 36(4) (1983): 518–534.

Armitage, Derek, Marschke, Melissa, and Plummer, Ryan. "Adaptive co-management and the paradox of learning." Global Environmental Change. 18(1) (2008): 86-98.

Western Historical Company. 1883. "Biographical Sketch of Mrs. Mary Phelps, Dec'd, Greene County, Missouri, Springfield". http://files.usgwarchives.net/mo/greene/bios/p4120007.txt. Accessed September 20, 2014.

Booth, Charles and Rowlinson, Michael. "Management and Organizational History: Prospects." Management & Organizational History. 1(1) (2006): 5-30.

Bosch, Ockert, Nguyen, Nam and Sun, Daowei "Addressing the critical need for a 'new way of thinking' in dealing with complex issues facing our societies." Business Systems Review. 2(2) (2013): 48-70.

Bower, Stephen."The Theology of the Battlefield: William Tecumseh Sherman and the U.S. Civil War."Journal of Military History. 64(4) (2000): 1005-1034.

Bucheli, Marcelo and Wadhwani, Dan. "A Return to the Past: History and Organization Studies." Aegis. 10(1) (2014): 5-13.

Carroll, Matthew, Blatner, Keith, Cohn,Patricia and Morgan, Todd "Managing Fire Danger in the Forests of the US Inland Northwest: A Classic 'Wicked Problem' in Public Land Policy." Journal of Forestry. 105(5) (2007): 239-244.

Cronon, William "The Uses of Environmental History." Environmental History Review. 17(3) (1993): 1-22.

Davison, Graeme. "Paradigms of public history." Australian Historical Studies. 24(97) (1993): 4-15.

Day, David, Fleenor, John, Atwater, Leanne, Sturm, Rachel and McKee, Rob. "Advances in leader and leadership development: A review of 25 years of research and theory." Leadership Quarterly. 25(1) (2014): 63-82.

De Stasio, Clotilde"Starving Vs Cramming: Children's Education and Upbringing in Charles Dickens and Herbert Spencer." Dickens Quarterly, 27(4) (2010): 299-306.

Degesys, Litauras. "Education as Liberation of the Self: Principles and Concepts of Learning and Teaching in the Pestalozzi Programme." In Josef Huber and Pascal Mompoint-Gaillard (eds), Teacher Education for Change: The Theory behind the Council of Europe Pestalozzi Programme, Council of Europe, 2011: 1-153

Donnachie, Ian. "Education in Robert Owen's New Society: The New Lanark Institute and Schools", In *Encyclopedia of Informal Education,* 2003, www.infed.org/thinkers/et-owen.html, accessed 20 August 2012.

Dovers, Stephen. "Sustainability and 'Pragmatic' Environmental History." Environmental History Review. 18(3) (1994): 21-36.

Dovers, Stephen. "Can Environmental History Engage with Policy." History Australia, 5(1) (2008): 3-2-3.6.

Dufty, Neil. "Engagement or Education?" Australian Journal of Emergency Management. 26 (3) (2011). https://ajem.infoservices.com.au/items/AJEM-26-03-08. Accessed April 13, 2012.

Fazey, Ioan, Fazey, John, Fischer, Joern et al. "Adaptive capacity and learning to learn as leverage for social-ecological resilience", Frontiers in Ecology and the Environment, 5 (2007): 1-7.

Finseth, Ian. "The Civil War Dead: Realism and the Problem of Anonymity", American Literary History, 25(3) (2013): 535-562.

Fischer, Joern, Dyball, Robert, Fazey, Ioan, Gross, Catherine, Dovers, Stephen, Ehrlich, Paul, Brulee, Robert, Christensen, Carleton and Borden, Richardet al. "Human Behavior and Sustainability", Frontiers in Ecology and the Environment, 10(3) (2012): 153-160. Borden, Richardet al. "Human Behavior and Sustainability", Frontiers in Ecology and the Environment, 10(3) (2012): 153-160.

Forrester, Jay. "System Dynamics, Systems Thinking, and Soft OR", Systems Dynamic Review, 10(2-3) (1994): 245-56.

"Goyder, George Woodroffe (1826-1898)," Australian Dictionary of Biography, National Centre of Biography, Australian National University. https://adb.anu.edu.au/biography/goyder-george-woodroffe-3647/text5683. Accessed May 1, 2009.

Griffiths, Tom. "The Humanities and an Environmentally Sustainable Australia", Australian Humanities Review, 43 (2007). http://australianhumanitiesreview.org/archive/Issue-December-2007/EcoHumanities/EcoGriffiths.html. Accessed January 15, 2010.

Griffiths, Tom."We Have Still Not Lived Long Enough", Inside Story, 16 (2009). http://insidestory.org.au/we-have-still-not-lived-long-enough/. Accessed April 30, 2010.

Handmer, John and Dovers, Stephen. "Policy Development and Design for Fire and Emergency Management", Australian Journal of Emergency Management, 23(1) (2008): 21-29.

Hilliard, David. "Unorthodox Christianity in South Australia", History Australia, 2(2) (2005): 38.1-38.10.

Hochstetler, Douglas. "Pedagogy and the Teacher Within: Reflections on Thoreau and Kinesiology in Higher Education." Quest, 65(2) (2013): 186-200.

Horn, Robert and Weber, Robert. "New Tools for Resolving Wicked Problems" (2007). https://www.strategykinetics.com/new_tools_for_resolving_wicked_problems.pdf. Accessed September 20, 2008.

Jones, Geoffrey and Wadhwani, Dohit. "Schumpeter's Plea: Rediscovering History and Relevance in the Study of Entrepreneurship." Harvard Business School Working Paper (2006): No. 06-036.

Jones, Geoffrey and Wadhwani, Daniel. "Entrepreneurship and Business History: Renewing the Research Agenda." Harvard Business School Working Paper (2006): No. 07-007.

Jones, Geoffrey, van Leeuwen, M.H.D. and Broadberry, Stephen. "The Future of Economic, Business, and Social History." Scandinavian Economic History Review, 60(3) (2012): 225-253.

Kipping, Matthias and Usdiken, Behlul. "History in Organization and Management Theory: More the Meets the Eye." The Academy of Management Annals, 8(1) (2014): 535-588.

Leclercq, Danielle. "The Pedagogical Foundations of the Pestalozzi Programme" In Josef Huber and Pascal Mompoint-Gaillard (eds), Teacher Education for Change: The Theory behind the Council of Europe Pestalozzi Programme, Strasbourg (2011): 1-153.

Lewis, James. "The Pinchot Family and the Battle to Establish American Forestry." Pennsylvania History. 66(2) (1999): 143-65.

Meadows, Donella. "Dancing with Systems", Systems Thinker, 13 (2002): 2-6.

Meehan, Sean. "Ecology and Imagination: Emerson, Thoreau, and the Nature of Metonymy", Criticism, 55(2) (2013): 299–329.

Miller, Char. "All in the Family: The Pinchots of Milford." Pennsylvania History, 66(2) (1999): 117-42.

Miller, Char. "Amateur Hour: Nathaniel H. Egleston and Professional Forestry in Post-Civil War America", Forest History Today, 25 (2005): 20-26.

Miller, Char. "Char Miller on Gifford Pinchot, Photographer." Environmental History, 8(2) (2003): 312-315.

Morris, Deidre. "Mueller, Sir Ferdinand Jakob Heinrich von (1825–1896)." in Australian Dictionary of Biography, National Centre of Biography, Canberra, Australian National University (1974). https://adb.anu.edu.au/biography/mueller-sir-ferdinand-jakob-heinrich-von-4266/text6893. Accessed April 28, 2014.

Neale, Greg. "Wanted: a 21st-century renaissance", Oxford Today, Hilary Issue (2010): 12-14.

Oelschlaeger, Max."The Roots of Preservation: Emerson, Thoreau, and the Hudson River School", http://nationalhumanitiescenter.org/tserve/nattrans/ntwilderness/essays/preserva.html. Accessed January 2013.

Ogilvy, Jay., "Systems Theory, Arrogant and Humble", World Futures, 69 (4-6) (2013): 332-344.

Olsson, Per, Falke, Carl and Berkes, Fikret. "Adaptive Comanagement for Building Resilience in Social–Ecological Systems." Environmental Management. 34(1) (2004): 75-90.

Reed, Mark, Evely, Anna, Cundill, Georgia et al. "What is Social Learning", Ecology and Society, 15(4) (2010): 1-10. http://www.ecologyandsociety.org/vol15/iss4/respl/. Accessed July 28, 2013.

Pawson, Eric and Dovers, Stephen. "Environmental History and the Challenges of Interdisciplinarity: An Antipodean Perspective", Environment and History, 9(1) (2003): 53-75.

Pittman, Nancy. "James Wallace Pinchot (1831 – 1908)." Yale Forestry and Environmental Studies News. Fall (1999): 4-7.

Pyne, Stephen. "Black Saturday: the Sequel." Peeling Back the Bark, (10 February 2009). https://fhsarchives.wordpress.com/2009/02/10/historian-stephen-j-pyne-on-the-australian-fires/. Accessed November 30, 2009.

Reese, William. "The Origins of Progressive Education." History of Education Quarterly, 41(1) (2001): 1-24.

Refshauge, Richard. "Brown, John Ednie (1848-1899)". Australian Dictionary of Biography, National Centre of Biography, Australian National University (1969). https://adb.anu.edu.au/biography/brown-john-ednie-3079. Accessed June 15 2012.

Remenyi, Daniel. "Tell Me a Story – a Way to Knowledge", The Electronic Journal of Business Research Methodology, 3(2) (2005): 133-140.

Richardson, Sarah. "A Day in the Life of a Soldier.", American History, 49(1) (2014): 44-50.

Rittel, Horst "Dilemmas in a General Theory of Planning", Policy Sciences, 4(2) (1973): 155-169.

Robin, Libby and Smith, Mike. "Australian Environmental History: Ten Years On", Environment and History, 14(2) (2008): 135–143.

Robinson, Rob and Johnston, Peter. "Forestry in South Australia." In *The Wakefield Companion to South Australian History*. (Adelaide: Wakefield Press, 2001).

Ronan, Kevin. "Solving 'wicked problems' linked to disasters, risk and uncertainty – children are truly our future", Australian Journal of Emergency Management, 29(1) (2014): 8-9.

Scharmer, O. "Uncovering the Blind Spot of Leadership", Leader to Leader, Winter (2008): 52-59.

Schumpeter, Joseph. "The Creative Response in Economic History", Journal of Economic History, 7 (1947): 149-159.

Senge, Peter. "Creating Schools for the Future Not the Past, For All Students", Leader to Leader, 65 (2012): 44-49.

Senge, Peter, Scharmer, Otto, Jaworski, Joseph and Flowers, Betty. "Awakening Faith in an Alternative Future", Reflections, 5(7) (2004): 1-11.

Suddaby, Roy, Hardy, Cynthia. and Huy,Quy. "Where are the New Theories of Organization?" Academy of Management Review, 36(2) (2011): 236-248.

Taylor, Andrew. "A Man is Conservative after Dinner: Ralph Waldo Emerson, Michel De Montaigne and the Appetites of Moderation", Forum for Modern Language Studies, 48(1) (2012): 1-18.

Tyrrell, Ian. "Acclimatisation and Environmental Renovation: Australian Perspectives on George Perkins Marsh", Environment and History, 10(2) (2004): 153-167.

Williams, Michael. "George Woodroffe Goyder: a Practical Geographer." Proceedings of the Royal Geographical Society of Australasia (S.A. Branch) Inc. (1978): 79-89.

Williams, Michael. "More and smaller is better: Australian rural settlement 1788-1914." In Michael Williams and Joseph Powell, *Australian Space Australian Time*. (Australia: Oxford University Press, 1975).

Williams, Michael."The Parkland Towns of Australia and New Zealand." Geographical Review, 56(1) (1966): 67-89.

Woolcock, Michael, Szreter, Simon, and Rao, Vijayendra. "How and Why Does History Matter for Development Policy?" Journal of Development Studies, 47(1) (2011): 70-96.

Zald, Mayer. "Spinning Disciplines: Critical Management Studies in the Context of the Transformation of Management Education." Organization, 9(3) (2002): 365–85.

Zaloom, Caitlin. "Mary Poovey: Interview." Public Culture. 24(1) (2012): 91.

Documentaries

Burns, Ken. "The National Parks: America's Best Idea." 2009. https://www.pbs.org/kenburns/the-national-parks/. Accessed May 21, 2010.

Websites

Burney Center, McGill University, https://www.mcgill.ca/burneycentre/. Accessed June 22 2014.

"Jardin des Plantes", http://www.britannica.com/EBchecked/topic/301392/Jardin-des-Plantes accessed 31 May 2010.

"Pinchot and Utilitarianism", https://www.fs.usda.gov/greatestgood/press/mediakit/facts/pinchot.shtml Accessed 30 June 2011.

Phillips Exeter Academy Bulletin. www.exeter.edu/documents/exeter_bulletin/spring_02/newground.html. Accessed June 20 2011.

"A Short History of Forestry in South Australia", http://www.pir.sa.gov.au/aghistory/left_nav/dept_of_agriculture_as_an_organisation/collaborating_organisations/department_of_woods_and_forests. Accessed April 30 2010.

"The US Civil War in Art", http://www.civilwarinart.org/items/show/93 accessed November 30 2013.

"William Brown, Canadian History", https://electricscotland.com/history/canada/brown_william.htm. Accessed June 23 2013.